NANOTECHNOLOGY SCIENCE AND TECHNOLOGY

INTRODUCTION OF FORENSIC NANOTECHNOLOGY AS FUTURE ARMOUR

NANOTECHNOLOGY SCIENCE AND TECHNOLOGY

Additional books and e-books in this series can be found on Nova's website under the Series tab.

NANOTECHNOLOGY SCIENCE AND TECHNOLOGY

INTRODUCTION OF FORENSIC NANOTECHNOLOGY AS FUTURE ARMOUR

RITESH KUMAR SHUKLA
AND
ALOK PANDYA
EDITORS

nova
science publishers
New York

Library of Congress Cataloging-in-Publication Data

ISBN: 978-1-53616-040-6

Published by Nova Science Publishers, Inc. † New York

CONTENTS

Preface		**vii**
Chapter 1	Introduction to Nanotechnology and Applications in Forensic Science *Hasmukh Gajera*	**1**
Chapter 2	Integration of Nanotechnology in Forensic Sciences *Neeti Kapoor, Ashish Badiye, Ritesh Kumar Shukla, Swati Dubey Mishra and Ankit Srivastava*	**33**
Chapter 3	Application of Nanotechnology in Latent Fingerprint Development *Paula Raijiwala, Ankit Srivastava and Ritesh Kumar Shukla*	**49**
Chapter 4	Nanosensors for Body Fluid Identification: Recent Advances and Future Opportunities *Anjani Prajapati and Alok Pandya*	**85**
Chapter 5	Explosives Detection and Nano Technique *Anand Lodha, Niha Ansari and Man Singh*	**115**

Chapter 6 Exploiting the Potential of DNA Fingerprinting in Forensic Science **145**
Gargi Bhattacharjee, Khushal Khambhati, Nisarg Gohil, Happy Panchasara, Shreya Patel and Vijai Singh

Chapter 7 Identification of Food Adulterants Using Nanotechnology **187**
Ramesh Chaudhari and Ashutosh Kumar

Chapter 8 Role of Nanotechnology in Forensic Document Examination and Preservation **215**
Alok Pandya, Hiteshkumar B. Roz and Ritesh Kumar Shukla

Chapter 9 Forensic Evidence: Cases, Legal Aspects and Aspects of Nanotechnology **227**
Heena Goswami and Aathama Kumar

Chapter 10 Nano-Tools for Illicit Drug Sensing: Advances and Challenges in Forensic Investigation **255**
Niha Ansari, Pinkesh Sutariya and Alok Pandya

About the Editors **291**

Index **295**

Related Nova Publications **307**

PREFACE

Forensic Science is an interdisciplinary subject that using principles and methodologies of various subjects including chemistry, biology, physics, geology, psychology, social science, engineering, help to law enforcement agencies to enforce criminal laws and regulations to resolve civil and criminal cases. Body fluids, fingerprints, footprints, illicit drugs, explosives etc. are the evidences that can be found at the scene of crime in trace amount. Foremost, to analyse these evidences first screening/ detection techniques are needed and thereafter, sophisticated instruments are required for analysis/confirmatory test. During these processes, chances to loss the amount of evidences are very high and even sometimes quantity of evidences are not adequate to meet the minimum requirement to analyse it. Generally, material scientist discussing new-fangled nanotechnology research and its conversation turns to the commercial aspects or potential issues around health, safety, and the environment. In recent years, some of them have shown their interest to integrate nanotechnology with basic sciences and applied for forensic purposes which are related to societal or community issues to change the way of examine the evidence.

Introduction of Forensic Nanotechnology as Future Armour is the first comprehensive book to consider both fundamental and applied aspects of Forensic nanotechnology. Forensic nanotechnology is an emerging field of the forensic science to investigate real-time crime scene and terrorist activities, inquiries, detecting the presence of explosive materials, biological indicators, document preservation, fingerprint enhancement and food adulteration using nanomaterial. Scientists and other researcher are working at nanotechnology applications that may steer the power of forensic investigation, but the challenges are to bring them from lab to the court. Moreover, some crucial concerns in forensic such as cost and time of analysis, effectiveness and wide availability of methods, accuracy and reliability of results are creating obstacles in forensic investigation and security.

In this book, as a unique feature of this work, we requested contributions to be submitted in a tutorial style, which means that state-of-the-art scientific content is enriched with each application of nanomaterials in forensic context. In this way, all chapters will indicate the exposure of this new area of research to solve the crime on site with the help of nanotechnology in various fields of forensic science such as detecting explosives, biological mediators, document preservation, fingerprint enhancement and food adulteration. The integration of laboratory processes onto a Nano platform is conceivably the most interesting advancement of nanotechnology highlighted in this book and some more issues like recent research developments, challenges and future opportunities are also addressed in this book.

The book is written for a wide readership, researchers, undergraduate and graduate students from diverse backgrounds such as chemistry, materials science and nanotechnology engineering, physics, life sciences, forensic science, and biomedical engineering. It can be used not only as a textbook but also as a review and reference book. However, the other many Nano technological applications for forensic analysis are yet to be studied where this book can be helpful to explore new opportunities. We hope that the chapters of this book will provide the reader with valuable insight as a revolutionary, protective tool in the fields of virtopsy, crime scene investigation, identification, forensic biology and toxicology. We also hope that after reading this book, researchers around the world will be motivated to enter into the field of forensic nanotechnology.

Dr. Ritesh Kumar Shukla
Dr. Alok Pandya

In: Introduction of Forensic Nanotechnology … ISBN: 978-1-53616-040-6
Editors: R. K. Shukla and A. Pandya

Chapter 1

INTRODUCTION TO NANOTECHNOLOGY AND APPLICATIONS IN FORENSIC SCIENCE

Hasmukh Gajera*

Department of Physical Sciences, Institute of Advanced Research, Gandhinagar, Gujarat, India

ABSTRACT

This chapter presents an introduction and history about Nanotechnology. The general synthesis approaches of nanomaterials are discussed briefly with different methods for preparation of nanomaterials. It also provides details of characterization of nanomaterials with different techniques and applications of nanotechnology and nanomaterials in the real world of forensic science and other field of science and technology.

Keywords: nano technology, synthesis, nanomaterials, characterization, applications and forensic science

---

* Corresponding Author's E-mail: hasmukh.gajera@iar.ac.in.

1. Introduction

In the past few decades, a tiny word paying enormous attention, interest and investigation around the world: **"nano"**. It has considerably changed every aspect of the way that we think in science and technology and will absolutely bring more and more surprises into our daily life as well as into the world in the future. Nanoscience and Nanotechnology based research has been rapidly increasing across the world during the last decade. It is accepted by the world scientific, industrial, government and business communities that nanotechnology will be an integral part in the development of future technologies. Nanotechnology will take part in an important position in the entire science and technological community.

The definitions of Nanoscience and Nanotechnology can be given as: Nanoscience is the study of phenomena and manipulation of material at atomic, molecular and macromolecular scales, where properties differ significantly from those at a larger scale. Nanotechnology is the design, characterization, production and applications of structures, devices and systems by controlling shape and size at the nanometer scale [1]. The nanometer scale is conventionally defined as 1 to 100 nm. One nanometer is one billionth of a meter (10^{-9} m). The size range is set normally to be minimum 1nm to avoid single atoms or very small groups of atoms being designated as nano objects. Therefore Nanoscience and Nanotechnologies deal with at least clusters of atoms of 1nm size as shown in Figure 1.

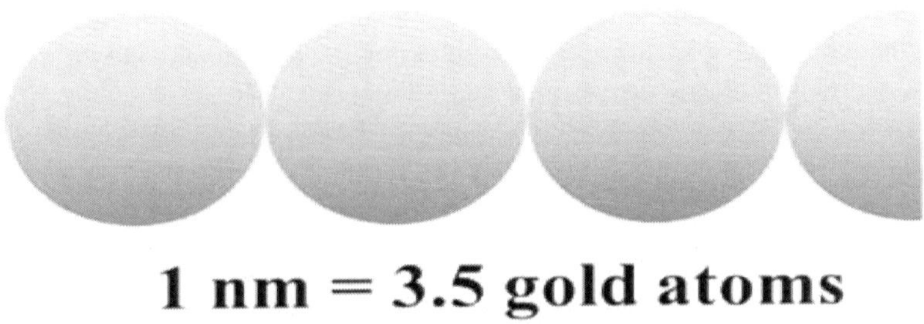

Figure 1. 3.5 gold atoms are equal to 1 nm.

Nanoscience is the scientific study at the atomic and molecular scale of molecular structures with at least one dimension measuring between 1 and 100 nanometers in order to understand their physico-chemical properties and identify ways to manufacture, manipulate, and control them. More generally, "Nanoscience" can mean research on any structure of matter that comprises of at least one nanometric dimension, without necessarily being aimed at the manufacture of objects or materials for a specific purpose. The results of certain studies on the behavior of ultrafine particles with dimensions of this scale (e.g., from diesel engine emissions) that are accidentally released into the air can be used in nanotechnology to better understand the behavior of nanoparticles designed and made in the laboratory. Therefore, the difference between Nanoscience and nanotechnology essentially reflects the conceptual distinction between science and technology. The word "nanotechnology" was first introduced in the late 1970s. While many definitions for nanotechnology exist, most groups use the National Nanotechnology Initiative (NNI) definition. The NNI calls something "nanotechnology" only if it involves all of the following: - Research and technology development at the atomic, molecular, or macromolecular levels, in the length scale of approximately 1 to 100 nanometer range. The comparison of an atom to bulk material is given in Figure 2. Creating and using structures, devices, and systems that have novel properties and functions because of their small and/or intermediate size. Ability to control and manipulation of matter at nanometer dimensions [2].

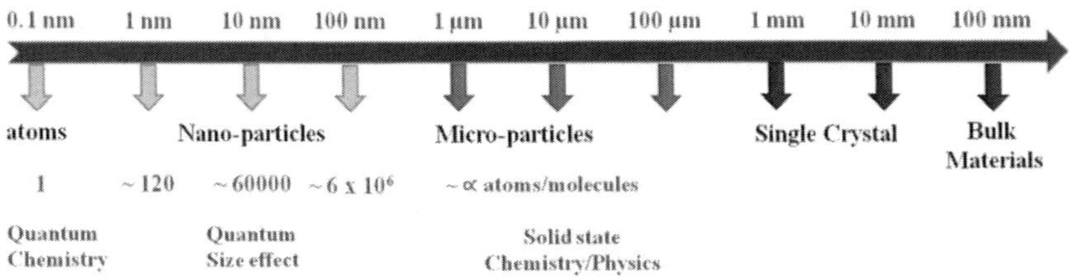

Figure 2. Size comparison of bulk material with atomic scale material.

In 1959 at the annual meeting a known physicist by the name Richard P Feynman quoted that there's plenty of room at the bottom [3]. In his introduction to the topic of his lecture he said, I would like to describe a field in which little has been done, but in which an enormous amount can be done in principle. This field will not tell us much of fundamental physics, but it might tell us much of great interest about strange phenomena that occur in complex situations. The term **"Nanotechnology"** was used first by the Japanese scientists Norio Taniguchi (1912-1999) in a 1974 paper on production technology that creates objects and features on the order of a nanometer. Inspired by Feynman's concepts, The American engineer K. Eric Drexler independently used the term "nanotechnology" in his 1986 book "Engines of Creation: The Coming Era of Nanotechnology", which proposed the idea of a nanoscale "assembler" which would be able to build a copy of itself and of other items of arbitrary complexity with atomic control. Also in 1986, Drexler co-founded The Foresight Institute (with which he is no longer affiliated) to help increase public awareness and understanding of nanotechnology concepts and implications. Thus, the emergence of nanotechnology as a field in the 1980s occurred through the convergence of Drexler's theoretical and public work, which developed and popularized a conceptual framework for nanotechnology, and high visibility experimental advances that drew additional wide scale attention to the prospects of atomic control of matter.

The invention of scanning tunneling microscope in the 1980s by IBM Zurich scientists and then the atomic force microscope allowed scientists to see materials at an unprecedented atomic level. The microscope's developers Gerd Binnig and Heinrich Rohrer at IBM Zurich Research Laboratory received a Nobel Prize in Physics in 1986 [4, 5].

Significant progress was obtained by IBM in 1990 when a team of physicists had spelled out the letters "IBM" using 35 individual atoms of xenon. Another breakthrough came in 1985 with the discovery of new shapes for molecules of carbon, known as the Bucky ball, which are round and consist of 60 carbon atoms [6]. This led to the discovery of a related molecular shape known as the carbon nanotube in 1991. Carbon nanotubes are other most promising areas of nanotechnology as they are about 100 times stronger than steel, but just a sixth of the weight; they have unusual heat and conductivity characteristics [7]. In parallel, studies of semiconductor nanocrystals led to the development of quantum dots, whose properties are between those of bulk semiconductors and discrete molecules. The major focus of research across the world continues to be researched on nanoscale properties, synthesis of materials and characterization, and application development to create useful devices and processes and reap economic benefits.

Nanotechnology is an interdisciplinary technology that has been booming in many areas during the recent decade, including material science, chemistry, forensic science, mechanics, electronics, optics, medicine, plastics, energy, electronics and aerospace. Further, nanotechnology was also expanded extensively to other fields due to the novel properties of nanomaterials discovered and to be discovered. For example, nanowires can be potentially used in nanophotonics, laser [8], nanoelectronics [9], solar cells [10], resonators [11] and high sensitivity sensors [12]. Nanoparticles can be potentially used in catalysts [13], functional coatings, nanoelectronics [14], energy storage [15], drug delivery [16] and biomedicines [17] and forensic toxicological analysis [18]. Nanostructured thin films can be used in light emitting devices, displays and high efficiency photovoltaics. These are only a limited part of the fast developing nanotechnology, yet numerous of other potential applications of nanomaterials have already been or will be discovered.

The field of nanotechnology, in general, is loosely divided into four subareas:

- Micro and nanoinstruments
- Nanoelectronics
- Nano biosystems and
- Nanoengineered materials

2. What Is Nano?

Nano is about as small as it gets in the world of regular chemistry, materials science, and biology. The word nano is about the phenomena that directly related to the size of systems with nanometer dimensions. Some of the unique aspects of nanosystems arise solely from the tiny (nano) size of the systems. The diameter of a hydrogen atom is about one-tenth of a nanometer, so the nanometer scale is the very smallest scale on which we might consider building machines and systems can be designed to exhibit novel and significantly improved physical, chemical and biological properties, phenomena and processes as a result of the limited size of their constituent particles or molecules.

The reason for such interesting and very useful behavior is that when characteristic structural features are intermediate in extent between isolated atoms and bulk macroscopic materials; i.e., in the range of about 10^{-9} m to 10^{-7} m (1 to 100 nm), the objects may display physical attributes substantially different from those displayed by either atoms or bulk materials.

The smaller something is, the larger its surface area is compared to its volume. This high surface-to-volume ratio is a very important characteristic of the system or matter at the nano scale. For example, imagine that you have a big block of ice with one-meter sides (see Figure 3). This block has a surface area of 6 square meters (1 square meter on side x 6 sides) and a volume of 1 cubic meter. In this case, the surface area to volume ratio for the ice block is 6/1 or 6. Suppose that cut the ice into 8 pieces that are one-half of a meter per side. The surface area of each piece of ice would be 1.5 square meters (0.5 m x 0.5 m x 6 sides). So the total surface area of all the pieces would be 12 square meters. However, the total volume of ice would stay the same: we haven't added or removed any ice. So in this case, the surface area to volume ratio is 12/1, or 12—twice the surface area to volume ratio of the block before it was cut. If you cut the ice into 27 pieces, the surface area increases to 18square meters, and the surface area to volume ratio is 18/1 or three times that of the uncut block. If you keep going, and cut the ice into 1000 small pieces, the surface area to volume ratio is 60/1 or ten times that of the uncut block!

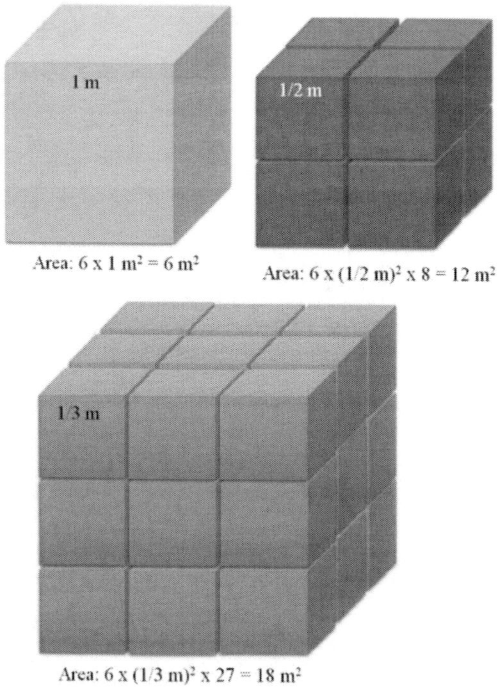

Figure 3. Total surface area increases as cut the block into smaller size, but the total volume stays constant.

The vastly increased ratio of surface area to volume makes interactions between the surfaces of particles very important. If something has more surface area, there are more places for other chemicals to bind or react with it. For example, fine powders offer greater reaction speed because of the increased surface area. Think about how much faster you can cool a glass of water if you put crushed ice in it rather than ice cubes. Nanoscale particles maximize surface area, and therefore maximize possible reactivity. The same concept is shown in Figure 4 where comparison between diameter of the Earth and diameter of Fullerene (C_{60}). The surface area to volume ratio of the earth near to 10^{-5} m^{-1} and of fullerene is near to10^9 m^{-1}.

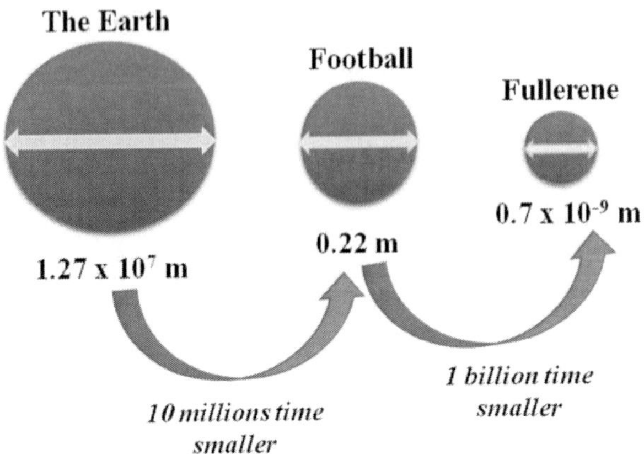

Figure 4. Total surface area increases as cut the block into smaller size, but the total volume stays constant.

3. Synthesis of Nanomaterials

3.1. General Approach for Synthesis

In general, top-down and bottom-up are the two main approaches for nanomaterials synthesis as shown in Figure 5.

(a) Top-down: size reduction from bulk materials.
(b) Bottom-up: material synthesis from the atomic level.

Top-down approaches are included in the typical solid –state processing of the materials start with bulk materials. This route is based with the bulk material and makes it smaller, thus breaking up larger particles by the use of physical processes like crushing, milling or grinding. Usually this route is not suitable for preparing uniformly shaped materials, and it is very difficult to realize very small particles even with high energy consumption. The biggest problem with this approach is the imperfection and irregulation of the surface structure. Such imperfection would have a significant impact on physical properties and surface chemistry of nanostructures and nanomaterials. It is well known that the conventional top-down technique can cause significant crystallographic damage to the processed patterns.

Bottom–up approaches refer to the build-up of a material from the bottom: atom-by-atom, molecule-by-molecule or cluster-by-cluster. This route is more often used for preparing most of the nano-scale materials with the ability to generate a uniform size, shape and distribution. It effectively covers chemical synthesis and precisely controlled the reaction to inhibit further particle growth. Although the bottom-up approach is nothing new, it plays an important role in the fabrication and processing of nanostructures and nanomaterials.

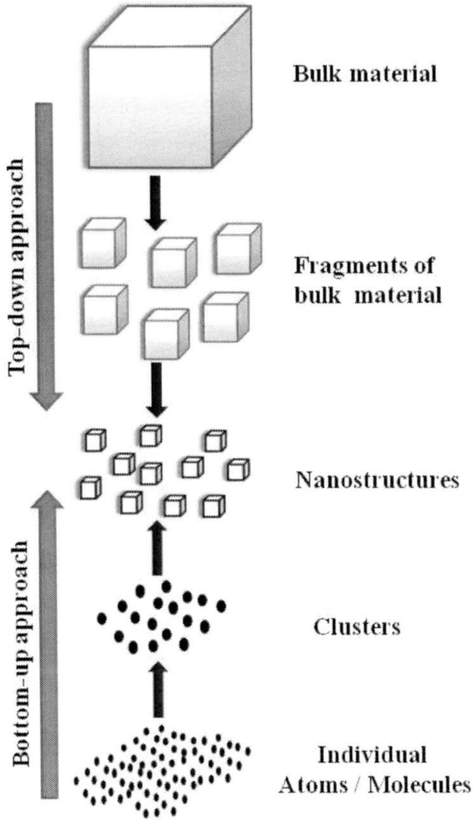

Figure 5. General approach for synthesis of nanomaterials by Top-down and Bottum-up.

In top-down fabrication a bulk material is reduced in size to nanoscale pattern. These seek to create smaller devices by using larger ones to direct their assembly.

3.1.1. Mechanical Milling/Ball Milling

A ball mill is a type of grinder used to grind materials into extremely fine powder for use in mineral dressing processes, paints, pyrotechnics, ceramics and selective laser sintering a 3D printing technology.

3.1.2. Chemical Etching

Chemical etching is the process of removing a layer of contamination on a metal or plastic surface through chemical erosion. It is mostly used on metals, though other materials are increasingly important. It was developed from armor-decorating, printing and etching processes developed during the Renaissance as alternatives to engraving on metal [19].

3.1.3. Laser Ablation

Laser ablation is the process of removing material from a solid (or occasionally liquid) surface by irradiating it with a laser beam.

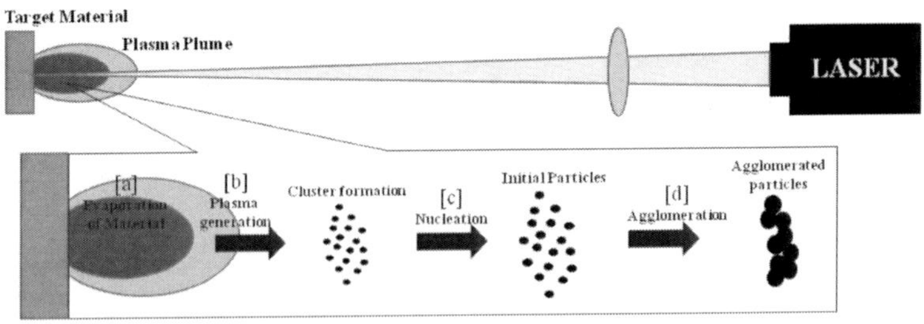

Figure 6. Schematic diagram of Laser ablation process.

At low laser flux, the material is heated by the absorbed laser energy and evaporates or sublimates. At high laser flux, the material is typically converted to plasma. Usually, laser ablation refers to removing material with a pulsed laser, but it is possible to ablate material with a continuous wave laser beam if the laser intensity is high enough. This process is used to synthesize nano particles of various metals and other materials which can vaporize under laser power [20]. The schematic representation of the steps involved in Laser ablation process is shown in Figure 6.

3.1.4. Sputtering

Sputtering is a process whereby atoms are ejected from a solid target material due to bombardment of the target by energetic particles. It only happens when the kinetic energy of the incoming particles is much higher than conventional thermal energies (>>1eV). This process can lead, during prolonged ion or plasma bombardment of a material, to significant erosion of materials, and can thus be harmful. On the other hand, it is commonly utilized for thin-film deposition.

The bottom-up techniques build or grow larger structures atom by atom or molecule by molecule. These techniques include

1. Chemical synthesis
2. Self-assembly
3. Positional assembly

3.1.5. Chemical/Electrochemical Method

3.1.5.1. Chemical Precipitation

Chemical precipitation method is a formation of a separable solid substance from a solution, either by converting the substance into an insoluble form or by changing the composition of the solvent to diminish the solubility of the substance in it. The distinction between precipitation and crystallization lies largely on whether emphasis is placed on the process by which the solubility is reduced or on that by which the structure of the solid substance becomes organized.

3.1.5.2. Electrochemical

The use of electrochemical methods represents an interesting option as many electrochemical and chemical reactions occur simultaneously when they are applied. Electrochemical treatment techniques have attracted a great deal of attention because of their versatility and environmental compatibility, which makes the treatments of liquids, gases, and solids possible. In fact, the main reagent is the electron, which is a "clean reagent".

3.1.6. Vapor Deposition Method

3.1.6.1. Chemical Vapor Deposition (CVD)

Chemical vapor deposition (CVD) is a chemical process used to produce high purity, high-performance solid materials.

The process is often used in the semiconductor industry to produce thin films. In typical CVD, the wafer (substrate) is exposed to one or more volatile precursors, which react and/or decompose on the substrate surface to produce the desired deposit. Frequently, volatile by-products are also produced, which are removed by gas flow through the reaction chamber. The schematic diagram of CVD setup is shown in Figure 7.

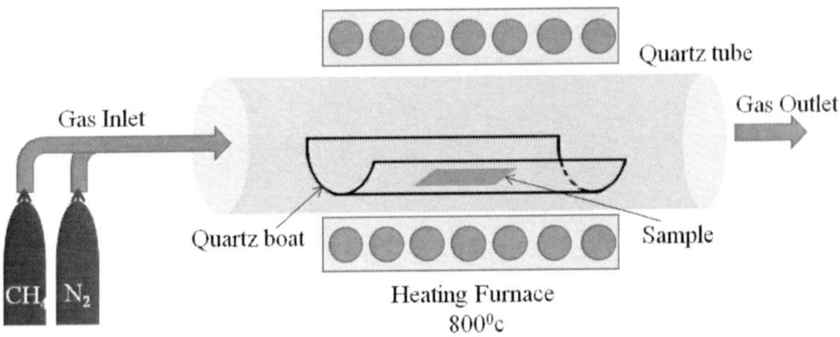

Figure 7. Schematic diagram of CVD setup.

3.1.6.2. Physical Vapor Deposition (PVD)

Physical vapor deposition (PVD) describes a variety of vacuum deposition methods used to deposit thin films by the condensation from a vaporized form of the desired film material onto various workpiece surfaces (e.g., onto semiconductor wafers). The coating method involves purely physical processes such as high temperature vacuum evaporation with subsequent condensation, or plasma sputter bombardment rather than involving a chemical reaction at the surface to be coated as in chemical vapor deposition.

3.1.7. Sol-Gel Processes

The sol-gel process is a method for producing solid materials from small molecules. The method is used for the fabrication of metal oxides, especially the oxides of silicon and titanium. The process involves conversion of monomers into a colloidal solution (Sol) that acts as the precursor for an integrated network (or Gel) of either discrete particles or network polymers. e) Spray pyrolysis is a process in which a thin film is deposited by spraying a solution on a heated surface, where the constituents react to form a chemical compound. The chemical reactants are selected such that the products other than the desired compound are volatile at the temperature of deposition. This process is particularly useful for the deposition of oxides and has long been a production method for applying transparent electrical conductor SnOx to glass. Also, spray pyrolysis is used for fabrication of semiconductor quantum dots. The schematic diagram of Sol Gel process is given as below in Figure 8 with all steps involved in process with possible final product.

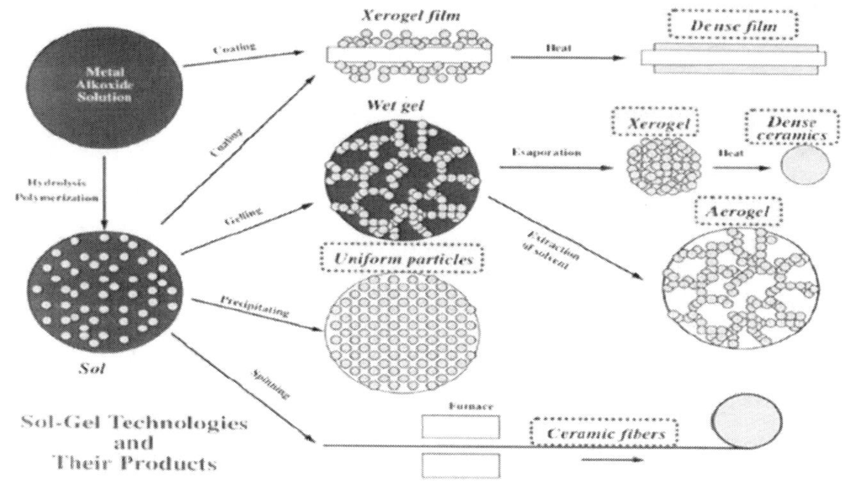

Figure 8. Sol-Gel process withsteps and possible final product. (https://www.nanowerk.com/how_nanoparticles_are_made.php).

3.1.8. Laser Pyrolysis

Laser pyrolysis, which is a gas phase process, has been pioneered and developed as a viable manufacturing platform (NPMTM) by NanoGram Corporation. It has been successfully employed in synthesizing various nanoscale materials like metals, simple and complex metal oxides, carbides, nitrides, rare-earth doped oxides, multi-elemental glasses among others. NPMTM process routinely produces very high-quality nanoscale particles with a high degree of crystallinity, generally spheroidal morphology and exceptionally tight particle size distributions.

4. Characterization Techniques for Nanomaterials

The techniques that can be used for the characterization of various nanomaterials comprise X-Ray diffraction, Scanning Electron Microscopy (SEM), Transmission Electron Microscopy (TEM), Atomic Force Microscopy (AFM), X-Ray florescence spectroscopy, UV-Vis spectroscopy and other spectroscopies. In this section, detail discussion is given on some of these characterization techniques.

4.1. X-Ray Diffraction

The German Physicist, Von Laue in 1912 was the first who took up the problem of X-ray diffraction (XRD) with the cause that, "if crystals were composed of regularly spaced atoms which might act as scattering centers for x-rays, and if X-rays were electromagnetic waves of wavelength about equal to the inter atomic distances in crystals, then it should be possible to diffract X-rays by means of crystals" [21]. Nowadays, X-ray diffractions most extensively used technique for the characterization of the materials. A lot of information can be extracted from the XRD data. This is an appropriate technique for all forms of samples, i.e., powder and bulk as well as thin film. Using this technique, one can get the information regarding the crystalline nature of a material, nature of the phase present, lattice parameter and grain size [22]. From the position and shape of the lines, one can obtain information regarding the unit cell parameters and micro structural parameters (grain size, micro strain, etc.), respectively. In case of thin films, the change in lattice parameter with respect to the bulk gives the idea about the nature of strain present in the system.

The interaction of X-ray radiation with crystalline sample is governed by Bragg's law, which depicts a relationship between the diffraction angles (Bragg angle), X-ray wavelength, and inter-planar spacing of the crystal planes. According to Bragg's law, the X-ray diffraction can be visualized as X-rays reflecting from a series of crystallographic planes as shown in Figure 9. The path difference between a pair of waves travelled through the neighboring crystallographic planes is determined by the inter-planar spacing. If the total path difference is equal to nλ (n being an integer), the constructive interference will occur and a group of diffraction peaks can be observed, which give rise to X-ray patterns. The quantitative account of Bragg's law can be expressed as:

$$2d_{hkl}\sin\theta = n\lambda$$

where d is the inter-planar spacing for a given set of *hkl* and θ the Bragg angle.

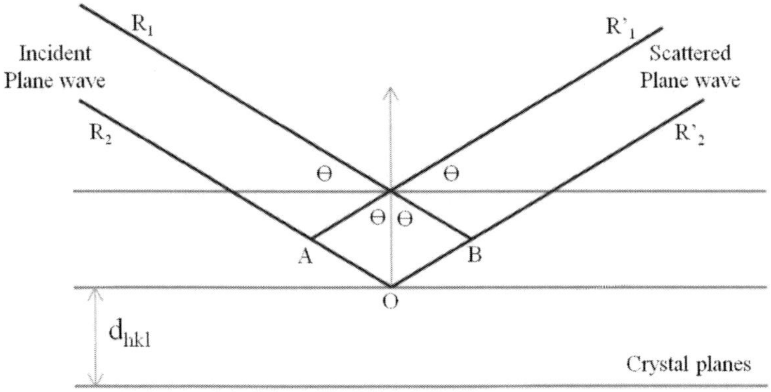

Figure 9. Geometrical illustrations of crystal planes and Bragg's law.

The diffraction method is based on the effect of broadening of diffraction reflections associated with the size of the particles (crystallites). All types of defects cause displacement of the atoms from the lattice sites. M.A. Krivoglaz in 1969 [23] derived an equation for the intensity of the Bragg reflections from a crystal defect, which enabled all the defects to be derived conventionally into two groups. The defects in the first group only lower the intensity of the diffraction reflections, but do not cause the reflection broadening. The broadening of the reflections is caused by the defects of the second group. These defects are micro-deformations, inhomogeneity (non-uniform composition of the substance over their volume) and small particle size. The size of nanomaterials can be derived from the peak broadening and can be calculated by using the Scherrer equation mentioned below, provided that the nanocrystallite size is less than 100nm.

$$D = \frac{k\lambda}{\beta \cos\theta}$$

where D is the average crystalline dimension perpendicular to the reflecting phases, **λ** the X-ray wavelength, ***k*** the Scherrer constant which equals 0.9 for spherical particles, whose value depends on the shape of the particle (crystalline, domain) and on diffraction reflection indices (hkl), and ***β*** is the full width at half maximum of the peaks. The Scherrer formula is quite satisfactory for small grains (large broadening) in the absence of significant microstrain.

4.2. Scanning Electron Microscopy (SEM)

Electron microscopes are scientific instruments that use a beam of energetic electrons to examine objects on a very fine scale. Electron microscopes were developed due to the limitations of Light Microscopes which are limited by the physics of light. In the early 1930s, this theoretical limit had been reached and there was a scientific desire to see the fine details of the interior structures of organic cells (nucleus, mitochondria..., etc.). This required 10,000X plus magnification which was not possible using existing optical microscopes.

The first Scanning electron microscope (SEM) debuted in 1938 (Von Ardenne) with the first commercial instruments around 1965. Its late development was due to the electronics involved in "scanning" the beam of electrons across the sample. Scanning electron microscopy (SEM) can provide a highly magnified image of the surface and the composition information of near surface regions of a material [24]. The resolution of the SEM can approach a few nanometers and the magnifications of SEM can be easily adjusted from about 10 times to 300,000 times.

The schematic diagram of SEM is given in Figure 10. In SEM, electron beam, accelerated by a relatively low voltage of 1-20 kV, is scanned on the specimen surface. As the electron beam strikes the surface, a large number of signals are generated from (or through) the surface in the form of electrons or photons.

These signals emitted from the specimen are collected by detectors to form images and the images are displayed on a cathode ray tube screen. There are three types of images produced in SEM: secondary electron images, backscattered electron images, and elemental X-ray maps. Secondary electrons (SE) are considered to be the electrons resulted from inelastic scattering with atomic electrons and with the energy less than 50 eV. The secondary emission of electrons from the specimen surface is usually confined to an area near the beam impact zone that permits images to be obtained at high resolution. These images, as seen on a cathode ray tube, provide a three dimensional appearance due to the large depth of field of the SEM as well as the shadow relief of the secondary electrons contrast. Backscattered electrons (BSE) are considered to be the electrons resulted from elastic scattering with the atomic nucleus and with the energy greater than 50 eV [24]. The backscattering will likely occur in a material of higher atomic number, so the contrast caused by elemental differences can be built up. After the primary electron beam collides with an atom in the specimen and ejects a core electron from the atom, the excited atom, then decays to its ground state and emit either a characteristic X-ray photon or an Auger electron. The energy dispersive X-ray detector (EDX) can sort the X-ray signal by energy and produce elemental images, so the spatial distribution of particular elements can be detected by SEM. SEM usually has a resolution of 1 nm for 1 KV, even resolution of 0.6 nm is possible for 5 KV.

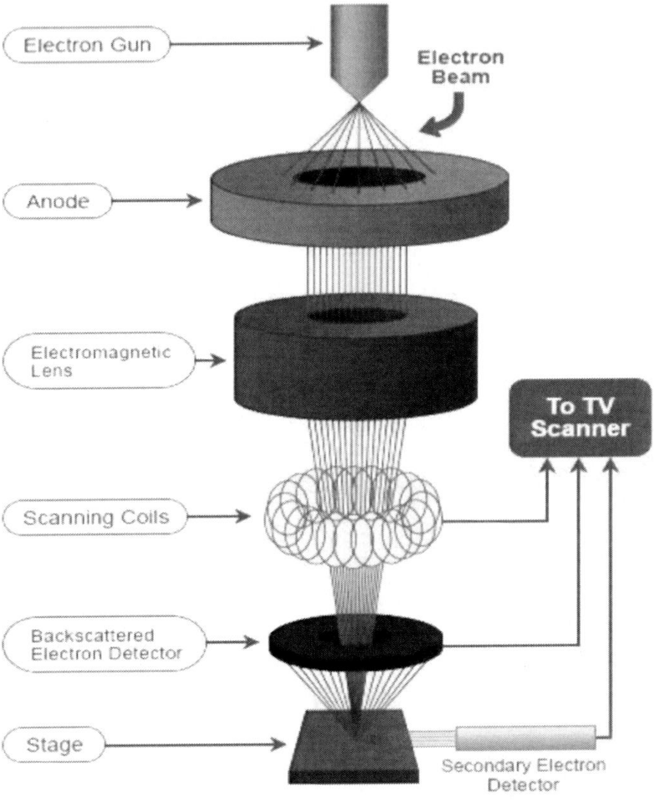

Figure 10. Schematic details of SEM (https://www.nanoimages.com/sem-technology-overview/).

4.3. Transmission Electron Microscopy (TEM)

Transmission electron microscopy (TEM) is a microscopy technique where a beam of electrons is transmitted through an ultra thin specimen, interacting with the specimen as it passes through. An image is formed from the interaction of the electrons transmitted through the specimen; the image is magnified and focused onto an imaging device, such as a fluorescent screen, on a layer of photographic film, or to be detected by a sensor such as a CCD camera. The Schematic diagram of TEM is shown in Figure 11. TEMs are capable of imaging at a significantly higher resolution than light microscopes, owing to the small de Broglie wavelength of electrons. This enables the instrument's user to examine fine detail- even as small as a single column of atoms, which is tens of thousands times smaller than the smallest resolvable object in a light microscope. TEM forms a major analysis method in a range of scientific fields, in both physical and biological sciences. TEMs find application in cancer research, virology, materials science as well as pollution and semiconductor research.

At smaller magnifications TEM image contrast is due to absorption of electrons in the material, due to the thickness and composition of the material. At higher magnifications complex wave interactions modulate the intensity of the image, requiring expert analysis of observed images. Alternate modes of use allow for the TEM to observe modulations in chemical identity, crystal orientation, electronic structure and sample induced electron phase shift as well as the regular absorption based imaging. The first TEM was built by Max Knoll and Ernst Ruska in 1931, with this group developing the first TEM with resolving power greater than that of light in 1933 and the first commercial TEM in 1939. Theoretically, the maximum resolution, d, that one can obtain with a light microscope has been limited by the wavelength of the photons that are being used to probe the sample, λ and the numerical aperture of the system, NA [25].

$$d = \frac{\lambda}{2n\sin\alpha} \approx \frac{\lambda}{2\,\mathrm{NA}}$$

Early twentieth century scientist's theorized ways of getting around the limitations of the relatively large wavelength of visible light (wavelengths of 400–700 nanometers) by using electrons. Like all matter, electrons have both wave and particle properties (as theorized by Louis-Victor de Broglie), and their wave like properties mean that a beam of electrons can be made to behave like a beam of electromagnetic radiation.

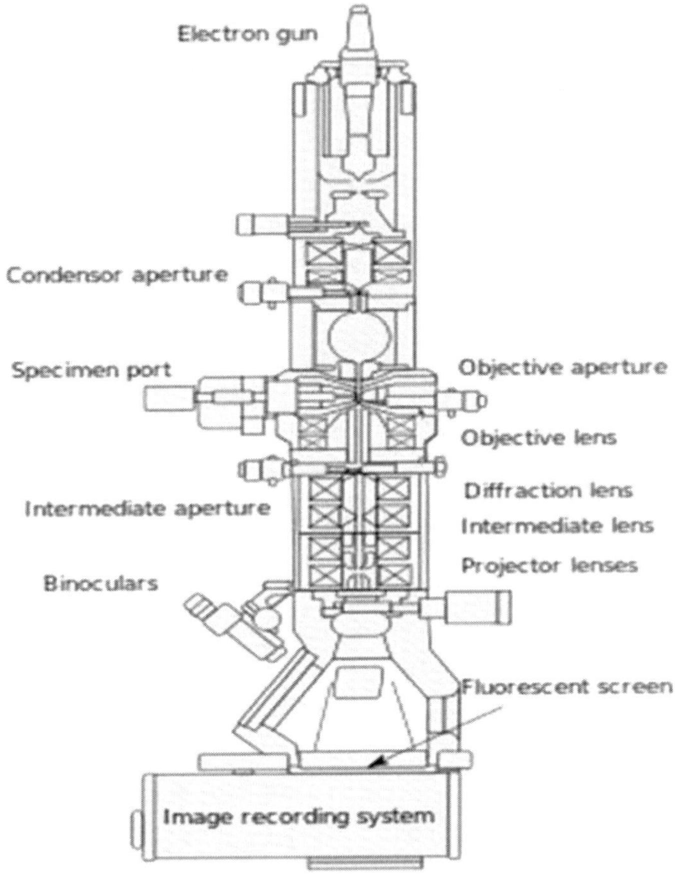

Figure 11. Schematic details of TEM.
(https://en.wikipedia.org/wiki/Transmission_electron_microscopy).

A TEM is composed of several components, which include a vacuum system in which the electrons travel an electron emission source for generation of the electron stream, a series of electromagnetic lenses, as well as electrostatic plates. The latter two allow the operator to guide and manipulate the beam as required. Also required is a device to allow the insertion into, motion within, and removal of specimens from the beam path. Imaging devices are subsequently used to create an image from the electrons that exit the system.

4.4. Atomic Force Microscopy (AFM)

The AFM, or scanning force microscope (SFM), is a very high-resolution type of scanning probe microscope, with demonstrated resolution of a fraction of a nanometer, more than 1000 times higher than the optical diffraction limit. The precursor to the AFM, the scanning tunneling microscope, was developed by Gerd Binnig and Heinrich Rohrer in the early 1980s, a development that earned them the Nobel Prize for Physics in 1986 [26]. The AFM is one of the foremost tools for imaging, measuring and manipulating matter at the nanoscale (Figure 12).

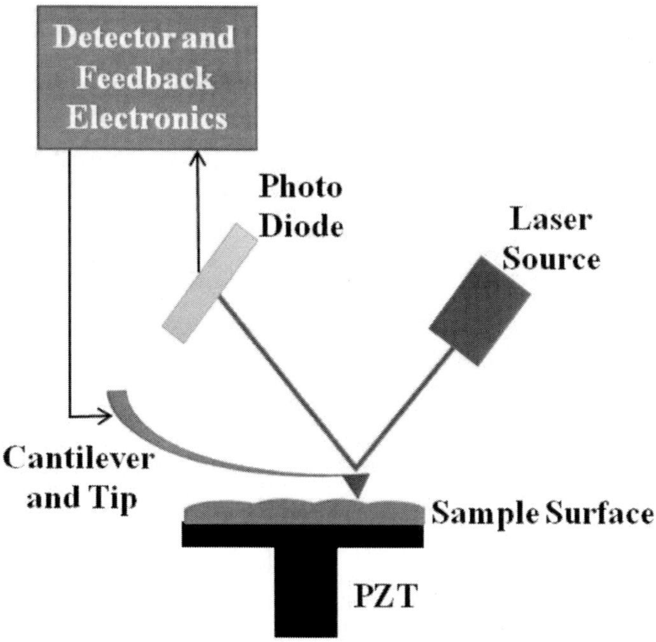

Figure 12. Block Diagram of Atomic Force microscope.

The information is adhered by "feeling" the surface with a microscale cantilever, a sharp tip (probe) at its end that is used to scan the specimen surface. The cantilever is typically silicon or silicon nitride with a tip radius on the order of several nanometers. Piezoelectric elements that facilitate tiny, but accurate and precise movement of the electronic command enable the AFM to have very precise scanning. When the tip is brought into proximity of a sample surface, the forces between the tip and the sample lead to a deflection of the cantilever according to Hooke's law. Depending upon the situation, forces that are measured by AFM include mechanical contact force, Van der Waals forces, capillary forces, chemical bonding, electrostatic forces, magnetic forces, Casimir forces, solvation forces, etc. In addition to these forces, other quantities could also be simultaneously measured through the use of a specialized type of probe.

Typically, the deflection is measured using a laser spot reflected from the top surface of the cantilever into an array of photodiodes. Other methods that are used include optical interferometry, capacitive sensing, or piezoresistive AFM cantilevers. These cantilevers are fabricated with piezoresistive elements that act as a strain gauge. Using a Wheatstone bridge, strain in the AFM cantilever due to deflection can be measured, but this method is not as sensitive as laser deflection or interferometry. If the tip was scanned at a constant height, a risk exists due to the possibility of the tip being damaged upon collision with the surface. Hence, in most cases a feedback mechanism is employed to adjust the tip-to-sample distance, and to maintain a constant force between the tip and the sample. Traditionally, the sample is mounted on a piezoelectric tube that can move the sample in the z direction for maintaining a constant force, and the x and y directions for scanning the sample. Alternatively, a "tripod" configuration of three piezo crystals may be employed, with each responsible for scanning in the x, y and z directions. This eliminates some of the distortion effects seen with a tube scanner. In newer designs, the tip is mounted on a vertical piezo scanner while the sample is being scanned in x and y using another piezo block. The resulting map of the area $S = f(x,y)$ represents the topography of the sample.

The AFM can be operated in a number of modes [27], depending on the application. In general, the possible imaging modes are divided into static (also called contact) modes and a variety of dynamic (or non-contact) modes [28] where the cantilever is vibrated.

4.5. UV-Vis Spectroscopy

Ultraviolet-visible (UV-vis) spectroscopy is widely utilized to quantitatively characterize organic and inorganic nanosized molecules. A sample is irradiated with electromagnetic waves in the ultraviolet and visible ranges and the absorbed light is analyzed through the resulting spectrum [29, 30]. The schematic diagram of working of UV-Vis spectrophotometer is shown in Figure 13. It can be employed to identify the constituents of a substance, determine their concentrations, and to identify functional groups in molecules. The samples can be either organic or inorganic, and may exist in gaseous, liquid or solid form. Different sized materials can be characterized, ranging from transition metal ions and small molecular weight organic molecules, whose diameters can be several Angstroms, to polymers, supramolecular assemblies, nano-particles and bulk materials.

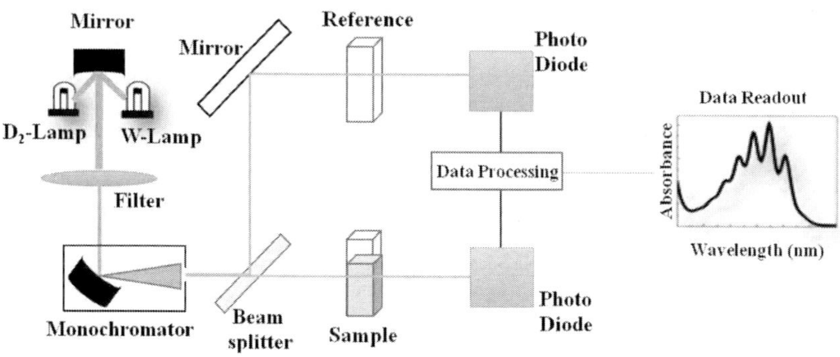

Figure 13. Schematic of UV- visible spectrophotometer.

Size dependant properties can also be observed in a UV-visible spectrum, particularly in the nano and atomic scales. These include peak broadening and shifts in the absorption wavelength. Many electronic properties, such as the band gap of a material, can also be determined by this technique. The energies associated with UV-visible ranges are sufficient to excite molecular electrons to higher energy orbitals [31]. Photons in the visible range have wavelengths between 800-400 nm, which corresponds to energies between 36 and 72 kcal/mol. A typical UV-vis absorption experiment setup is shown schematically in Figure 8 for a liquid sample. A beam of monochromatic light is split into two beams, one of them is passed through the sample, and the other passes a reference.

After transmission through the sample and reference, the two beams are directed back to the detectors where they are compared. The difference between the signals is the basis of the measurement. Liquid samples are usually contained in a cell (called a cuvette) that has flat, fused quartz faces. Quartz is commonly used as it is transparent to both UV and visible lights. UV-vis spectroscopy offers a relatively straightforward and effective way for quantitatively characterizing both organic and inorganic nanomaterials.

Furthermore, as it operates on the principle of absorption of photons that promotes the molecule to an excited state; it is an ideal technique for determining the electronic properties of nanomaterials. In the spectrum of nanoparticles, the absorption peak's width strongly depends on the chemical composition and the particle size. As a result, their spectrum is different from their bulk counterparts. An important consequence of using the UV-vis spectroscopy is that the band gap of nanosized materials can be determined.

5. Applications of Nanotechnology in Forensic Science

In the field of forensic science for the investigation, different types of nanomaterials and nanotechnologies have been used so far for various applications like development of fingerprints, trace evidence, examination of bloodstain, drug analysis, Forensic DNA Typing, estimation of the time of death, used as Nano trackers for determining the authenticity of the items and also have lots of possibilities to explore more possible applications of nanotechnology in this field as well as in other fields also.

5.1. Nanotechnology and Fingerprints

The use of nanoparticles has recently shown a great potential in producing the next generation of fingerprint development techniques known as nano fingerprints [32]. Finger prints may be latent, patent or plastic in nature. Since the fingerprint is a complex mixture of natural secretions of the body (mostly sweat from a different type of glands) and contaminations from the environment [33].Currently, avariety of techniques are being used to develop the latent fingerprints such as iodine fuming, ninhydrin, silver nitrate and cyanoacrylate. However, in the case of additionally aged prints the environment might contaminant the latent fingerprints, to such environments, the fingerprints may suffer from reduced reactivity to detection from these methods. Therefore, over the past few years, nano-scientist have been using increasingly sophisticated techniques to exploit the effects of nanoparticles such as gold, cadmium sulphide (CdS) and zinc sulphide (ZnS) for the development of latent fingerprints in order to better identify the fingerprints [34].Thus Nanotechnology offers the opportunity to develop new fingerprint techniques with superior properties including; improved selectivity, improved contrast with the background and increased sensitivity [32].

5.2. Nanotechnology and Trace Evidence

Detection and analysis of trace evidence are one of the important aspects of any crime scene investigation in forensic science. Atomic force microscopy (AFM) is a nanotechnology tool that can be used to generate forensic information on a various range of trace evidence including the measurement of cuticle step heights in hair, the morphological and elastic properties of hairs and fibers, and the density with structural properties of bone and enamel [32].According to a recent study on the fiber, it was reported that AFM is a very powerful tool in the forensic examination of fiber evidence due to its capability to distinguish between different environmental exposures or forced damages to fibers [35].

5.3. Nanotechnology and Examination of Bloodstain

The examination of blood stains in forensic science plays a major role in investigation during any crime scene. There are lots of techniques are available for the detection and identification of blood spots for examination of suspected bloodstains like as tetramethylbenzidine, phenolphthalein can be used. By using these types of techniques, there is a change in color when they come into contact with hemoglobin or peroxides in the blood. In recent time, the problem with this is to determine the age of the blood spot and difficult to get by using normal analysis method [36]. Since Atomic Force Microscopy (AFM) is a fast adopting tool into the examination of the age of blood stains, potentially providing legal medical information for the forensic investigations. So that it is important to test the elasticity of blood by recording force-distance curves by using AFM. The elasticity pattern decreases over time, which is the most critical process of bloodstain during the drying and coagulation of blood sample.

5.4. Forensic DNA Typing

In recent times the quantification of post-polymerase chain reaction (post-PCR) is the most effective application of forensic nanotechnology for the micro fluidic systems. In a very short period of time, i.e., within 30 minutes DNA samples can be quantified even in the nanoliter range using commercially available Agilent 2100 bio analyzer. These systems are presently being used to quantify mitochondrial DNA in many forensic science laboratories. Due to their small size, the prospective of such devices being used at the scene of crime is extensively specified [37]. Nanoparticles have received considerable attention in biology and medicine because of their promising applications. Pan et al. [38] apprehended hot start-like effect in gold nanoparticle-based PCR, which has promising applications in nano biological and biomedical studies. DNA extracted from urine using organic reagents while carboxylated magnetic nano particles are used as solid phase adsorbents to isolate intact DNA for PCR amplification [39].

5.5. Nanotechnology and Estimation of the Time of Death

In the medical field investigation of time since death (TSD) is a very important and vital step. A precise evaluation of TSD estimation may be help full to provide the accurate time and cause of death. There are various parameters through which TSD can be estimated in commonly used methods. These parameters included algor mortis, changes in the eye, postmortem hypostasis, rigor mortis, changes in decomposition, stomach and bowels content, contents of urinary bladder, and anecdotal evidences. All of these parameters can only provide an approximate time of death. All of these parameters can only provide an approximate time of death [40]. Out of these body fluids, the vitreous humor (VH) remains unchanged even after a long time span of death. It exhibited that the biochemical changes (level of amino acid) in VH are slow and therefore, by the help of this biochemical analysis, TSD can be estimated accurately. Recently, a smart, rapid, sensitive, cost-effective, and lab on chip method have been developed that provides easy determination of cysteine, an amino acid [41]. This method may lead to estimate the TSD up to 96 h till cystine concentration from VH increases significantly and showing a linear correlation with expansion of TSD. In the near future, this method can be developing a micro fluidic system, smart approach-based detection to estimate the cysteine concentration and correlate it with TSD estimation [42].In the near future, amino acid concentration from VH can be detected by fluorescent nanoparticles and then quantification could be assessed using flow cytometry which would be a possible approach to estimate TSD precisely [40,41].

5.6. Used as Nano Trackers

With advanced in technology, barcoding and tracking systems are promising tools to combat the crome. Trackers and barcodes are being used as a preventing tool in the form of a secret signature on the products to keep away from people from thieving. Trackers can also be used for security purposes, as they can be used to avert prisoners from absconding. In this case, nano trackers are injected into the prisoners' body which helps to track them if they escape. Nano trackers are also enabled to monitor the convicts after being released. Nano trackers help to track them easily if they are again involved in a crime [43].With the help of nanotechnology, various tools are being developed that could be able to differentiate forged products from authentic one. In this way, nanotechnology helps to prevent counterfeit crime. Nanotechnology has been developed nanofibers and nanodots with diverse colors to detect duplicate products. These nanofibers and nanodots have been detected using various forensic light sources that are meant to decipher the nanofibers and nanodots. So, in this way, counterfeit crime has been prevented [44]. Nanotechnology is also relevant in those cases in which petroleum products are involved such as arson cases to examine the petroleum hydrocarbon traces. Furthermore, nanotechnology is also pertinent in the cases related to cyber (mobile and internet) and sting operations (authentication of tapes and identification of speakers). In these cases, the investigators scrutinize suspected tapes by identifying the coating of nanomaterials on its surface which is a marking of authentication [45].

References

[1] Pradeep, T. *Nano: The Essentials – Understanding Nanoscience and Nanotechnology*, Tata McGraw Hill Publishing Co. Ltd., Chennai, 2008.

[2] Rao, CNR; Muller, A; Cheetham, AK. *The chemistry of nanomaterials*., 2004, 1(1). Weinheim: Wiley-VCH. 2004.

[3] Feynman, Richard P. There's Plenty of Room at the Bottom. *Engineering and Science*., 1960, 23(5), 22-36.

[4] Binnig, G; Rohrer, H. *IBM Journal of Research and Development*, 1986, (30), 355-369.

[5] Robinson, AL. Electron microscope inventors share Nobel Physics prize. *Science*., 1986, 234, 821-823.

[6] Kroto, HW. *Symmetry, space, starts and C.* Nobel Lecture., 1996, December, 7.

[7] Iijima, S. Helical microtubules of graphitic carbon. *Nature*., 1991, 354(6348), 56-58.

[8] Zimmler, MA; Bao, J; Capasso, F; Müller, S; Ronning, C. Laser action in nanowires: Observation of the transition from amplified spontaneous emission to laser oscillation. *Applied Physics Letters*., 2008, 93(5), 051101-3.

[9] Javey, A; Nam, S; Friedman, RS; Yan, H; Lieber, CM. Layer-by-layer assembly of nanowires for three-dimensional, multifunctional electronics. *Nano letters*., 2007, 7(3), 773-777.

[10] Castelletto, S; Boretti, A. Noble metal nanoparticles in thin film solar cells. *Nanoscience and Nanotechnology Letters*, 2013, 5(1), 36-40.

[11] Tanner, SM; Gray, JM; Rogers, CT; Bertness, KA; Sanford, NA. High-Q GaN nanowire resonators and oscillators. *Applied Physics Letters*, 2007, 91(20), 203117.

[12] Fan, Z; Lu, JG. Gate-refreshable nanowire chemical sensors. *Applied Physics Letters*, 2005, 86(12), 123510.

[13] Haruta, M; Tsubota, S; Kobayashi, T; Kageyama, H; Genet, MJ; Delmon, B. Low-temperature oxidation of CO over gold supported on TiO2, α-Fe2O3, and Co3O4. *Journal of Catalysis*, 1993, 144(1), 175-192.

[14] Rau, NJ; Welles, EA; Wenthold, PG. Anionic substituent control of the electronic structure of aromatic nitrenes. *Journal of the American Chemical Society*, 2013, 135(2), 683-690.

[15] Wang, J; Polleux, J; Lim, J; Dunn, B. Pseudocapacitive contributions to electrochemical energy storage in TiO2 (anatase) nanoparticles. *The Journal of Physical Chemistry C.*, 2007, 111(40), 14925-14931.

[16] Mah, C; Zolotukhin, I; Fraites, TJ; Dobson, J; Batich, C; Byrne, BJ. Microsphere-mediated delivery of recombinant AAV vectors in vitro and in vivo. *Mol Ther.*, 2000,1, S239.

[17] Pereiro, M; Baldomir, D; Botana, J; Arias, JE; Warda, K; Wojtczak, L. Biomedical applications of small silver clusters. *Journal of Applied Physics.*, 2008, 103(7), 07A315.

[18] Redasani, VK; et al. Development and Validation of Spectroscopic Methods for the Estimation of Mupirocin Calcium in Bulk and in Ointment Formulation. *J Forensic Toxicol Pharmacol.*, 2015, 4(2), 139.

[19] Li, X; Bohn, PW. Metal-assisted chemical etching in HF/H2O2 produces porous silicon. *Applied Physics Letters.*, 2000, 77(16), 2572-2574.

[20] Myungjoon, K; Saho, O; Taesung, K; Hidenori, H; Takafumi, S. Synthesis of Nanoparticles by Laser Ablation: A review. *Powder and Particle Journal.*, 2007, 34, 80-90.

[21] Eckert, M. Max von Laue and the discovery of X-ray diffraction in 1912. *Annalen der Physik.*, 2012, 524(5), A83-A85.

[22] Liu, D; Wang, Q; Chang, HLM; Chen, H. Variant structure in metal-organic-chemical-vapor-deposition-derived SnO2 thin films on sapphire (0001). *Journal of Materials Research.*, 1995, 10(6), 1516-1522.

[23] Krivoglaz, MA; Azároff, LV.Theory of X-Ray and Thermal Neutron Scattering by Real Crystals. *Physics Today.*, 1970,23, 78.

[24] Brundle, CR; Evans, CA; Wilson, S. *Encyclopedia of materials characterization: surfaces, interfaces, thin films.* Gulf Professional Publishing., 1992, 70-84.

[25] Fultz, B; Howe, JM. *Transmission electron microscopy and diffractometry of materials.* Springer Science & Business Media., 2012, 105-107.

[26] Binnig, G; Quate, CF; Gerber, C. Atomic force microscope. *Physical review letters*,1986, 56(9), 930-933.

[27] Elings, VB; Magonov, S. *Digital Instruments Inc. Method for improving the operation of oscillating mode atomic force microscopes*. U.S. Patent, 6,008,489, 1999.

[28] Yang, CW; Hwang, S; Chen, YF; Chang, CS; Tsai, DP. Imaging of soft matter with tapping-mode atomic force microscopy and non-contact-mode atomic force microscopy. *Nanotechnology*., 2007, 18(8), 084009-16.

[29] Clark, BJ; Frost, T; Russell, MA. eds. *UV Spectroscopy: Techniques, instrumentation and data handling*. Springer Science & Business Media., 1993, 4, 7-10.

[30] Perkampus, H\H. *Recent developments in UV-VIS spectroscopy: InUV-VIS Spectroscopy and Its Applications*. Springer, Berlin, Heidelberg., 1992, 81-130.

[31] Atkins, PW. Chemisorbed and Physisorbed Species, *A Textbook of Physical Chemistry*, University Press Oxford: NY, 936, (1980).

[32] Chauhan, V; Singh, V; Tiwari, A. Applications of nanotechnology in forensic investigation. *Int. J. Life. Sci. Scienti. Res*, 2017, 3(3), 1047-1051.

[33] Champod, C; Lennard, CJ; Margot, P; Stoilovic, M. *Fingerprints and other ridge skin impressions*. CRC press., 2016, 17-29.

[34] Wood, MA. *A novel approach to latent fingermark detection using aptamer-based reagents*., 2014, Doctoral dissertation.

[35] Chen, YF.Forensic applications of nanotechnology. *Journal of the Chinese Chemical Society*, 2011, 58(6), 828-835.

[36] Strasser, S; Zink, A; Kada, G; Hinterdorfer, P; Peschel, O; Heckl, WM; Nerlich, AG; Thalhammer, S. Age determination of blood spots in forensic medicine by force spectroscopy. *Forensic Science International*, 2007, 170(1), 8-14.

[37] Endres, RG; Cox, DL; Singh, RR. Colloquium: The quest for high-conductance DNA. *Reviews of Modern Physics*., 2004, 76(1), 195-214.

[38] Dun, P; Yanqin, W; Lijuan, M; Chunhai, F; Jun, H. Nanomaterials-based Polymerase Chain Reactions for DNA Detection. *Current Organic Chemistry.*, 2011, 15(4), 486-497.

[39] Shan, Z; Zhou, Z; Chen, H; Zhang, Z; Zhou, Y; et al. PCR-ready human DNA extraction from urine samples using magnetic nanoparticles. *J Chromatogr B Analyt Technol Biomed Life Sci.,* 2012, 15(881-882), 63-68.

[40] Pandya, A; Shukla, RK. New perspective of nanotechnology: role in preventive forensic. *Egyptian Journal of Forensic Sciences.*, 2018, 8(57), 1-11.

[41] Swann, L; Forbes, S; Lewis, S. Analytical separations of mammali and ecomposition products for forensic science: a review. *Anal Chem Act.*, 2010, 682, 9–22.

[42] Ansari, N; Lodha, A; Menon, S. Smart platform for the time since deathdetermination from vitreous humor cystine. *Biosens Bioelectron.*, 2016, 86, 115-121.

[43] Hogard, E; Ellis, R. *Community safety. Innovation and Evaluation,* Chester, UK: Chester Academic Press, 2007.

[44] Teeuw, WB; Vedder, AH; Custers, BHM; Dorbeck-Jung, BR; Faber, ECC; Iacob, SM; Koops, BJ; Leenes, RE; de Poot, HJG; Rip, A; Vudisa, JN. *Security applications for converging technologies, impact on the constitutional state and the legal order.* Boom Juridische Uitgevers, Den Haag, 2008.

[45] Chalmers, J; Edwards, H; Hargreaves, M. *Infrared and Raman spectroscopy in forensic science*. Wiley, Singapore, 2012.

In: Introduction of Forensic Nanotechnology … ISBN: 978-1-53616-040-6
Editors: R. K. Shukla and A. Pandya © 2019 Nova Science Publishers, Inc.

Chapter 2

INTEGRATION OF NANOTECHNOLOGY IN FORENSIC SCIENCES

Neeti Kapoor*[1]*, Ashish Badiye*[1,*]*, Ritesh Kumar Shukla*[2]*,
***Swati Dubey Mishra*[3] *and Ankit Srivastava*[4]**

[1]Department of Forensic Science, Government Institute of Forensic Science, Nagpur, Maharashtra, India
[2]Biological and Life Sciences, School of Arts and Sciences, Ahmedabad University, Ahmedabad, Gujarat, India
[3]Shri Vaishnav Institute of Forensic Science, Shri Vaishnav Vidyapeeth Vishwavidyalaya, Indore, Madhya Pradesh, India
[4]Dr. APJ Abdul Kalam Institute of Forensic Science and Criminology, Bundelkhand University, Jhansi, Uttar Pradesh, India

ABSTRACT

Nanotechnology is a rapid growing field of science & Technology which leads to appear in several areas of forensic science including security documents, detection of drugs, paints, inks, explosive and latent fingerprint development. In addition, various type of nano-based sensors

* Corresponding Author's E-mail: badiye.ashish@gmail.com.

are also developing which can shift the paradigm of forensic research from laboratory based detection system to on field detection system.

Precisely, in this chapter we attempted to elucidate the current status of nanotechnology, specifically nanoparticles and nano-structured particle-based techniques & their several uses. In particular, this chapter focused on the applications and limitations of techniques relying on metal-containing nanoparticles and nano-structured particles. Proper development and application of nanoparticles specifically for forensic uses will revolutionized the modern Crime Investigations. But, continuous and high throughput research is urgently needed in the field of forensic nanotechnology.

Keyword: nanotechnology, forensic nanotechnology, nano-based sensor

1. Introduction

Nanotechnology and Nanoparticles are the buzz words in the modern world. It is deemed that one day "Nano" will change the world as we know it today. The era of "bigger the better" seems to have paved way for "lesser the better." Nanotechnology is ready to take the world by a storm. It has found its way in all the routine aspects of our life. It has shown promising potential in varied fields like robotics, biomechanics, pharma, biomedical, optics, biosensor, forensics, etc. Let us get an insight into this "nano" world.

Nanotechnology is the manipulation or self-assembly of individual atoms, molecules, or molecular clusters into structures to create materials and devices with new or vastly different properties. This can be achieved by reducing the size of the smallest structures to the nanoscale (e.g., photonics applications in nanoelectronics and nanoengineering) or by manipulating individual atoms and molecules into nanostructures, which more closely resembles chemistry or biology. The definition of nanotechnology is based on the prefix "nano," which is from the Greek word meaning "dwarf". In more technical terms, the word "nano" means 10-9, or one billionth of something. To illustrate this, a virus is approximately 100 nanometres (nm) in size. Nanotechnology opens a completely new world of opportunities and solutions in all kinds of areas.

An example for daily use is copying the water and dirt-repelling effect of leafs of the Lotus flower, and to use it for applications like nano-coating of mobile phones and car surfaces to make them splash resistant/water resitant. Apart from the field of diagnostics and analytics, nanotechnology is already appearing in the textile industry, the energy sector, electronics and automotive industry, to name just a few [1].

The nanotechnology field is a fast-growing research field [2]. Nanoparticles are particles with at least one dimension less than 100 nm. Nanotechnology is a collective term that implies the capacity to work with materials at a nanometer scale [3]. The word 'nano' means small which represents the size of a particle that falls in the order of 10^{-9} m and is one billionth of a meter. Nanoparticles have high effectiveness in a range of applications across disciplines of knowledge. Nanotechnology involves imaging, measuring, modeling and manipulating matter at nano scale. It is a multidisciplinary scientific undertaking, involving creation and utilization of materials, devices, or systems on the nanometer scale and is currently undergoing explosive development that is sufficient to bring about revolution in many disciplines of science [4].

In nanotechnology, a *particle* is defined as a small object that behaves as a whole unit in terms of its transport and properties. Furthermore, according to size they may be subclassified as *Fine particles*, *Ultrafine particles, Nano-particles, Nano-clusters, Nano-powders* and *Nano-crystals.* In terms of diameter, fine particles cover a range between 100 and 2500 nanometers, while ultra-fine particles, are sized between 1 and 100 nanometers. Similar to ultra-fine particles, *nanoparticles* are sized between 1 and 100 nanometers [5]. Although the size of most molecules would fit into the above outline, individual molecules are usually not referred to as nanoparticles. Nanoclusters have at least one dimension between 1 and 10 nanometers and a narrow size distribution. Nano-powders are agglomerates of ultra-fine particles, nanoparticles, or nanoclusters [6]. Nanometer-sized single crystals, or single-domain ultra-fine particles, are often referred to as nanocrystals. Nanoparticles may or may not exhibit size-related properties that differ significantly from those observed in fine particles or bulk materials [7, 8, 9]. Nanoparticles are effectively a bridge between bulk materials and atomic or molecular structures [10].

The key qualities of nanoparticles are their shape (including aspect ratios where appropriate), size, and the morphological sub-structure of the substance. Nanoparticles are presented either as an aerosol (mostly solid or liquid phase in air), a suspension (mostly solid in liquids) or an emulsion (two liquid phases). In the presence of chemical agents (surfactants), the surface and interfacial properties may be modified [11]. Indirectly such agents can stabilise against coagulation or aggregation by conserving particle charge and by modifying the outermost layer of the particle.

The stimulating and sometimes unanticipated properties of nanoparticles are partially owed to the characteristics of the surface of the material which dominates the properties in lieu of the bulk properties. Nanoparticles display a number of distinct properties relative to bulk material. For example, the bending of bulk copper (wire, ribbon, etc.) occurs with movement of copper atoms/clusters at about the 50 nm scale. Nanoparticles often have surprising visible properties because they are small enough to confine their electrons and produce quantum effects. For example, *gold nanoparticles* appear deep red to black in solution [12].

Nanoparticles have a very high surface area to volume ratio. This provides a tremendous driving force for diffusion, especially at elevated temperatures. Sintering can take place at lower temperatures, over shorter time scales than for larger particles [10]. This theoretically does not affect the density of the final product, though flow difficulties and the tendency of nanoparticles to agglomerate complicates matters. The large surface area to volume ratio also reduces the incipient melting temperature of nanoparticles.

These properties are well documented in the literature but new ones are being discovered at a rapid pace. Each of these properties are being used and utilised in a number of ways which were unthought of and unheard of earlier. The potiental of growth of the field of nanoparticles is enormous.

2. Recent Advancements in Forensic Science

Nanoparticles appear in several areas of forensic science including security documents, paints, inks, and reagents that develop latent prints (Figure 1). One reagent (known as the silver physical developer) that visualizes the water insoluble components of latent print residue is based on the formation of highly charged silver nanoparticles.

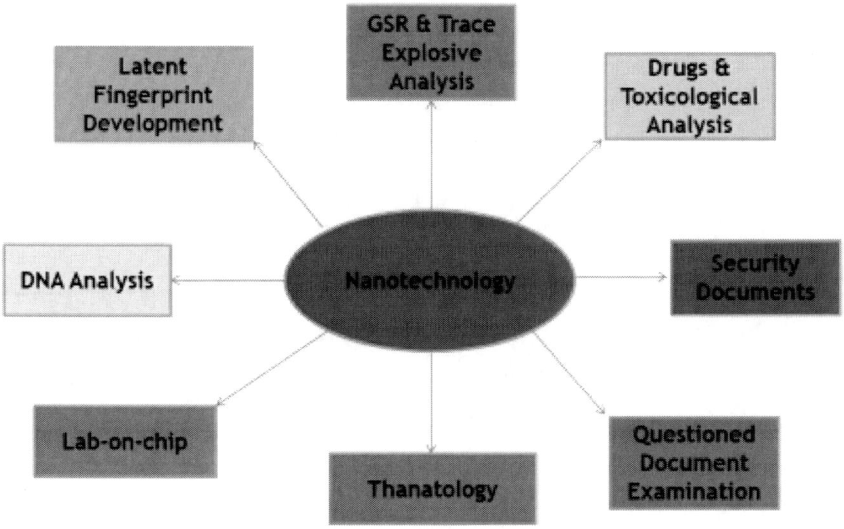

Figure 1. Application of nanotechnology on forensic science.

These attach to and grow on the residue and generate a silver image. Another such reagent involves highly charged gold nanoparticles. These attach to the residue forming a weak gold image which can be amplified with a silver physical developer. Nanoparticles are also used in items such as paints, printing inks, and writing inks. Paints and most printing inks consist of nano-sized pigments in a vehicle. However, certain modern ink jet printing inks now contain nano-sized pigments to improve their light fastness and most gel inks are also based on nano scale pigments. These nanoparticles containing materials often appear as evidence and are thus subject to forensic characterization. Both luminescent (quantum dots), up-converting nano scale phosphors, and non luminescent nanoparticles are used as security tags to label product, add security to documents, and as anti-counterfeiting measures. These assist in determining whether an item is genuine or fraudulently made.

2.1. In Fingerprint Development

Latent fingerprint (LFP) detection is an important task in forensic science [13]. It is a simple and effective means for the identification of individuals. A specific application of nanotechnology is its use *to improve fingerprint recovery*. Latent fingerprints are formed by a chemical secretion from the fingertips. Nanoparticles or quantum dots could be designed to bind to these secretions, revealing the fingerprint in greater detail, and increasing the chance of matching it. Researchers have developed sol gel particles which use fluorescent dyes such as Texas red, increasing the quality of the recovered fingerprint.

Development of nanomaterials which maximize the surface interaction with endogenous substances on the ridges to enhance the contrast of the fingerprints is an important application of nanotechnology in LFP detection. However, most developments in this area have mainly focused on the visualization of the physical pattern of the fingerprints and failed to explore the molecular information embedded in LFPs.

The application of *metal-containing nanoparticles and nano-structured particles* for fingermark detection is attracting significant interest as advances in nanoscience are being incorporated into the field of forensic fingermark detection. Some researchers have integrated certain distinctive properties of gold nanoparticles (AuNPs) with imaging mass spectrometry for both the visualization and molecular imaging of LFPs [14]. Two contrasting colors (blue and pink), arising from different surface plasmon resonance (SPR) bands of the AuNPs, reveal the optical images of LFPs. The laser desorption/ionization property of the AuNPs allows the direct analysis of endogenous and exogenous compounds embedded in LFPs and imaging their distributions without disturbing the fingerprint patterns [15]. The simultaneous visualization of LFP and the recording of its molecular images not only provide evidence on individual identity but also resolve overlapping fingerprints and detect hazardous substances. Nano carbon powder offers a green chemistry approch for the development of LFP. It have several advantages over the traditional black powder.

Recently, photophysical properties of carbogenic nanoparticles (C- dots) has found to have many uses. Carbogenically coated Silica particles has a promising forensic importance. These particles provide an effective approach for the LFP and also offers contrast against patterned and multicoloured backgrounds [16].

Detection of latent fingerprints using lipophilic and polycationic polymer chitosan has been explored by a few. The gold nanoparticle deposition on chitosan treated latent fingerprints enhances contrast, making the fingerprint identification possible. Chitosan being the second most abundant natural polymer, this technique can be an inexpensive and efficient method for fingerprint enhancement and its subsequent detection. This simple technique has a potential of immediate application in forensic science[17].

Researchers have tested a fingerprint powder based on zinc oxide nanoparticles on surfaces such as glass, polyethylene and aluminium. Accordingly, the 20-nanometer zinc oxide particles clump together in one micrometer sized flower-like crystals, much smaller than the current 10 micrometer fingerprint powder particles. It was found that illuminating the nanoparticles with ultraviolet light caused them to fluoresce without the addition of any fluorescent dyes. It was also observed that the zinc oxide nanoparticle powder gave a much clearer picture of the fingerprints, compared to conventional powders. “When you dust with a powder, you're hoping that it will stick only to the fingerprint, but often it will stick to everything, The nanoparticles are very good at sticking to the fingerprint residue but not to the background surface,” The system worked exceptionally well in wet conditions [18].

Very recently, a group of Indian researchers have developed a new nanomaterial to overcome difficulties in deciphering invisible and often unclear fingerprints accidentally left behind by criminals are often difficult to decipher because available techniques can't capture all the nuances in them. The nanomaterial, consisting of semiconductor particles that are a million times smaller than a millimeter, is luminescent. The new material has been developed by doping manganese and copper atoms on zinc sulphide nanosystem. By replacing zinc atoms with those of copper and manganese, researchers could change optical properties of zinc sulphide nanosystem. When this material is applied on latent fingerprints, its strong visible luminescence property helps decipher fingerprints. All one will need to do is sprinkle the nanomaterial on surface where fingerprint has been left behind, capture the image using UV lamp and process it to reconstruct the fingerprint. Researchers said that such fingerprints could be captured using a mobile phone too and transmitted to forensic lab from any crime scene [19].

Although more research is needed before some of the techniques existing can be implemented in routine casework, nanotechnology is likely to play a major role in the future to deliver more selective and more sensitive ways to detect and enhance fingermarks [20].

Detection of drugs and their metabolites in dusted latent fingermarks by mass spectrometry - A hydrophobic silica dusting agent containing carbon black has been used with latent finger marks to demonstrate that the agent can act as an enhancing matrix to generate a simple method for detecting a range of drugs using surface assisted laser desorption/ionisation time-of-flight mass spectrometry (SALDI-TOF-MS) in positive ion reflectron mode [21]. The dusting agent produces developed marks for locating/visualizing the prints whilst also acting as a SALDI-TOF-MS enhancer that is equivalent to the standard matrix enhancer 2,5-dihydroxybenzoic acid. This method has been applied to the analysis of latent fingermarks for contact residues on fingers, and for detection of illicit drugs for both parent drugs and their metabolites. Analysis was performed by direct MS analysis of the pre-dusted fingermarks on the surface of a target plate and, following lifting using commercial tape, MS analysis of the lifted marks. The presence of the parent drug and its metabolites could be confirmed using SALDI-TOF-MS-MS following high energy collision induced dissociation when characteristic and unique fragmentation patterns were observed in each case. The distribution of these compounds on fingermarks was subsequently demonstrated using commercially available imaging software.

Direct detection of nicotine and cotinine in dusted latent fingermarks of smokers by using hydrophobic silica particles and MS- Sub-micron sized hydrophobic silica particles doped with carbon black have been employed with latent fingermarks on glass and metal surfaces to develop a simple method for detecting nicotine and cotinine using surface assisted laser desorption/ionisation-time of flight-mass spectrometry (SALDI-TOF-MS) in positive ion reflectron mode. Dusting of surfaces enables location of marks for conventional identification of details but the particles also act as a laser desorption/ionisation enhancing agent (LDI) equivalent to a standard matrix enhancer 2,5-dihydroxybenzoic acid used in matrix assisted LDI-TOF-MS (MALDI-TOF-MS). The method has been applied in the analysis of smokers' latent fingermarks on metal and glass surfaces. The metal surfaces were analysed following direct MS analysis of the pre-dusted prints, while the glass surfaces were analysed following lifting using commercial tape, and then MS analysis of these pre-dusted prints. The presence of nicotine and cotinine in smokers' marks was confirmed using SALDI-TOF-MS-MS following high energy collision-induced dissociation when characteristic fragmentation patterns were observed for each compound[22].

2.2. Other Applications

Researchers have discovered a method of making *the brightest ever synthesized fluorescent silica* particles [23]. The fluorescence of these particles is 170 times brighter than any particles of similar size created so far. These nanostructured macroscopic silica particles have potential applications in medicine, forensic science and environmental protection, among many other uses. There are a multitude of applications for these microscopic "flashlights." These particles, which are less than 1/10th the size of a human hair, could be used like holograms to prove a product's authenticity, imbedded in various objects or fabric for tracking, or used like an "invisible ink" to prove that someone touched an object. They could be used like a barcode, encoded and deciphered not in terms of light, but in terms of colors. Using only commercially available dyes, we could create and track about 100 trillion combinations of these color barcodes. Because these particles are so bright, it is possible to detect even a single colored particle easily [23].

Nanotechnologies allow the development of new classes of sensors which can better answer current security constraints. Nanotubes, nanoparticles, and quantum dots enable sensors to be further miniaturized, become more sensitive, and to be used directly in the field rather than the lab. Such sensors can be highly selective while simultaneously detecting various hazardous agents. Two classes of sensors have been identified based on direct or indirect detection. It appears that nanotechnology applications for sensors are essentially for the detection of molecules or organisms, i.e., biological or chemical agent detection [24].

Nanotechnology has been pronounced as a *preventative forensic* tool. It has several advantages over the traditional analytical techniques and devices that are being used for forensic purposes. Nanotechnology has potential to significantly contribute in the time since death estimation, analysis of durg facilliated crimes etc. Importance of tools like nanoprobes, nanodevices and nanochips has been underlined[25].

DNA electrochemical sensors have application in forensics as they are cheap, simple, reliable, sensitive and selective for genetic identification. The human disease-related mutated genes or DNA sequence detection at low cost can be verified by the electrochemical-based biosensor. A range of different chemistries is used by the DNA-based electrochemical biosensors, out of which the interactions of nanoscale material with recognition layer and a solid electrode surface are most interesting. A diversity of advancements has been made in the field of electrochemical detection. A number of electrochemical methods have been reported in the literature based on impedance of voltametry. Gold nanoparticle based electrochemical chip for DNA identification has been reported. Electrochemical assays based on tracer quantum dots of ZnS, CdS, PbS, and CuS have also been reported. A metal sulphide (Cadmium/Lead) based detection of DNA provides atomolar sensitivity of the targeted DNA [26].

Researchers have developed fluorescent dye-doped silica nanoparticles functionalized with oligonucleotides as labels for chip-based sandwich DNA assays. The nanoparticles are composed of a silica matrix that encapsulates large numbers of fluorophores and can detect DNA target down to 1fM. They have also used similar particles to detect single bacterium cells by modification of the fluorescent nanoparticles with specific monoclonal antibodies [27].

Detection of Chemical, Biological, Radiological, Nuclear, Explosives - This application area specifically considers approaches to the detection of harmful substances associated with a security threat, such as a chemical or biological agent. Given the tsunamic rise in instances of terrorism globally, this is an area of high interest, particularly for governments, and there is a wide range of research work being carried out. Nanotechnology offers the possibility to make smaller, more sensitive and integrated detection platforms for each of these substances. These could typically be used in transit points like airports, train stations, seaports and borders to prevent the transport of such materials. Nanotechnology has a significant contribution in the detection of trace explosives.

Detection of Biological Substances - Work being carried out in the healthcare industry on the detection of pathogens, toxins and other substances has applications in civil security. One of the primary events triggering concern about the weaponization of biological substances were the anthrax attacks of 2001, in which anthrax spores were sent by mail to US media outlets and senators, resulting in the deaths of five people. Nanoparticles have shown exceptional colorimetric properties that can easily replace traditional fluorescent detection systems.

Nanotechnology-enabled detection methods include the use of metallic nanowires, to which antibodies corresponding to specific pathogens are attached. Fluorescent antibodies are then added, which bind to any pathogens which are present. Measurement of the fluorescence then indicates the presence and concentration of pathogens.

A sensor developed by NASA uses an array of vertically aligned carbon nanotubes, each tipped with a probe molecule. When the probe molecule comes into contact with a target substance, an electrical impulse is generated. A variety of different probing molecules can be used, enabling a single array to detect multiple substances. Further, Carbon nanotube research may pave ways for smaller, faster computers, networks, and sensors [28].

Detection of Radioactive and Nuclear Material - A detection technology which nanotechnology may improve is the use of scintillators; materials which emit photons when exposed to radiation. Materials including zinc oxide nanoparticles may work as scintillators but demonstrate improved energy resolution.

Detection of Explosives – Nanotechnology can contribute towards making of better explosives detection systems, which may be able to go deep and sniff out the hidden or mixed or camoflauged explosives. This woud be crucial at public transportation hubs like airports, railway stations, subways, metro stations, buses, etc. It would be also useful at the cargo hubs wherein several thousand cargo containers needs to be continuosly monitored for the presence of unwanted dangerous substances such explosives and drugs. It would also find application in clandestine laboratories or hideouts of potential terrorists, arsonists, illegal explosives factories and godowns.

Questioned document analysis is another field which warrants the use of nanoparticles to resolve key questions like the age of the document, age of the ink, etc. that still remains unaswered whenever they arise during routine criminal and civil investigations.

Anti-Conterfeiting- It is imperative to prevent counterfeiting by incorporating various anti-counterfeiting measures. Nano-particles may be used for the prevention of counterfeiting of important security documents, currency notes, legal documents, drugs and pharmaceutical products, to name a few. Nanobarcodes can be used for covert security in inks, adhesives, laminates, paper, packaging, and films.

CONCLUSION

Nanoparticles are the future, not only of Forensic Science but of the whole world. Nanoparticles and nanotechnology are here to stay. Everything is and will be affected by the development of nanoparticles. They will continue to prove to be very useful in our daily life apart from the various applied branches and fields of science and technology.

In the broad scientific arena, applications of nanotechnology seem to be only limited by human imagination. Nano-mites and Nano-bots may be a reality soon. They may be injected in the body of a person and may have different applications ranging from tracking a person, to detection and extraction of poison from the body, to repairing of the broken bones and injuries, etc. The nano-bots may be useful in retrieval of information from damaged harddisks, mobile phones, laptops, etc. Preventive as well as proactive applications are numerous and needs to be worked upon.

In particular, it is generally considered that nanotechnology may lead to the development of new materials and reagents with superior characteristics to conventional ones. Nanotechnology provides new opportunities in surface-based science. It is a leading contender to become the 'entirely different approach' in being able to develop the science further. Nanotechnology has proved to be a promising area of research.

Nanoscience and nanotechnology be it nanosensors, nanotubes, nanowires, quantum dots, nanodots, nanopaper, nanocomposites, nanofibers, nanocrystals, nanocrystalline alloys, nanofluids, nanostructured solutions, magnetic nanoparticles, nanobarcodes, nanospray, nanospheres, nanobots, smart nanoparticles, etc. are generating an impact on forensic studies. This chapter briefly reviewed the current status of nanotechnology, specifically nanoparticles and nano-structured particle-based techniques & their several uses in forensic science. Proper development and application of nanoparticles specifically for forensic uses will revolutionized the modern crime investigations. Some of these methods are not yet sufficiently mature for routine implementation in casework. Continuous and largescale focused research is required to continue the development of nanotechnology-based detection methods to optimize the results obtained from this approach.

REFERENCES

[1] https://www.nanowerk.com/nanotechnology/reports/reportpdf/report72.pdf.

[2] Ostiguy C., et al., *Health Effects of Nanoparticles* - Second Edition, Studies and Research Projects/Report R-589, Montréal, IRSST, October 2008, 114 pages.

[3] Sachindri Rana and P. T. Kalaichelvan, (2013). "Ecotoxicity of Nanoparticles," *ISRN Toxicology,* Article ID 574648, 11 pages. https://doi.org/10.1155/2013/574648.

[4] Retrieved from http://shodhganga.inflibnet.ac.in/bitstream/10603/36092/8/10_chapter1.pdf.

[5] Goodsell, David S. (2004). *Bionanotechnology: Lessons from Nature*, 1st edition Wiley-Liss; 1 edition.

[6] Fahlman, B. D. (2007). *Materials Chemistry*. Springer. pp. 282–283.

[7] Buzea, C.; Pacheco, I. I.; Robbie, K. (2007). "Nanomaterials and nanoparticles: Sources and toxicity." *Biointerphases*. 2 (4): MR17MR71. doi:10.1116/1.2815690.

[8] ASTM E 2456 – 06 *Standard Terminology Relating to Nanotechnology*.

[9] Valenti G, Rampazzo R, Bonacchi S, Petrizza L, Marcaccio M, Montalti M, Prodi L, Paolucci F (2016). "Variable Doping Induces Mechanism Swapping in Electrogenerated Chemiluminescence of Ru(bpy)32+ Core–Shell Silica Nanoparticles." *J. Am. Chem. Soc.* 138 (49): 15935–15942. doi:10.1021/jacs.6b08239.PMID 27960352.

[10] Retrieved from *https://en.wikipedia.org/wiki/Nanoparticle*.

[11] Scenihr/002/05 (Scenihr- Scientific Committee on Emerging and Newly Identified Health Risks modified Opinion (after public consultation) on: *The appropriateness of existing methodologies to*

assess the potential risks associated with engineered and adventitious products of nanotechnologies).

[12] Retrieved from https://www.sciencedaily.com/terms/nano particle.htm.

[13] Retrieved from https://pubs.rsc.org/en/content/articlelanding/2016/ay/c6ay01811d.

[14] Retrieved from http://hub.hku.hk/handle/10722/68844.

[15] Retrieved from https://www.ncbi.nlm.nih.gov/pmc/articles/PMC3709732/.

[16] Fernandes, Diogo & Krysmann, Marta & Kelarakis, Antonios. (2016). Carbogenically coated silica nanoparticles and their forensic applications. *Chem. Commun.*. 52. 10.1039/C6CC02556K.

[17] Ul Islam, Naveed & F. Ahmed, Kazi & Sugunan, Abhilash & Dutta, Joydeep. (2007). Forensic Fingerprint Enhancement Using Bioadhesive Chitosan and Gold Nanoparticles. *Proceedings of the 2nd IEEE International Conference on Nano/Micro Engineered and Molecular Systems*, IEEE NEMS 2007. 10.1109/NEMS.2007.352047.

[18] Retrieved from http://www.abc.net.au/science/articles/2008/07/07/2292298.htm

[19] Retrieved from https:// www.downtoearth.org.in/ news/ science-technology/ now- a- technology- that- can- make- reading- latent-fingerprints-faster-62844

[20] Retrieved from https://www.nanowerk.com/spotlight/spotid=6291.php.

[21] Retrieved from https://pubs.rsc.org/en/content/articlehtml/2009/an/b813957c.

[22] Benton, M., Rowell, F., Sundar, L. and Jan, M. (2010), Direct detection of nicotine and cotinine in dusted latent fingermarks of smokers by using hydrophobic silica particles and MS. *Surf. Interface Anal.*, 42: 378-385. doi:10.1002/sia.3112

[23] Retrieved from http://internal.clarkson.edu/news/2007/news-release_2007-03-30-4.html.

[24] Retrieved from https://www.nanowerk.com/nanotechnology/reports/reportpdf/report72.pdf.

[25] Pandya, A. & Shukla, R.K. *Egypt J Forensic Sci* (2018) 8: 57. https://doi.org/10.1186/s41935-018-0088-0.

[26] Bushra Rafique, Mudassir Iqbal, Tahir Mehmood, Muhammad Ashraf Shaheen, (2019) "Electrochemical DNA biosensors: a review", *Sensor Review*, Vol. 39 Issue: 1, pp.34-50, https://doi.org/10.1108/SR-08-2017-0156

[27] Tan et al., (2003). "Ultrasensitive DNA Detection Using Highly Fluorescent Bioconjucated Nanoparticles" *Journal of American Chemical Society* 125 pp. 11474-11475.

[28] Retrieved from https:// www.nasa.gov/ centers/ ames/ research/ technology-onepagers/carbon_nanotubes_vertical.html

In: Introduction of Forensic Nanotechnology … ISBN: 978-1-53616-040-6
Editors: R. K. Shukla and A. Pandya © 2019 Nova Science Publishers, Inc.

Chapter 3

APPLICATION OF NANOTECHNOLOGY IN LATENT FINGERPRINT DEVELOPMENT

***Paula Raijiwala*[1], *Ankit Srivastava*[2] *and Ritesh Kumar Shukla*[1,*]**

[1]Biological and Life Sciences, School of Arts and Sciences, Ahmedabad University, Ahmedabad, Gujarat, India
[2]Dr. APJ Abdul Kalam Institute of Forensic Science & Criminology, Bundelkhand University, Jhansi, Uttar Pradesh, India

ABSTRACT

Nanotechnology is the most promising and emerging technology, which provides great benefit to science and society. Nanotechnology also shifts the paradigm of forensic research and nowadays is applied in paint, ink, Gunshot residue, security document and enhancement of latent fingerprint. Latent fingerprints are the most commonly found physical evidence at the scene of the crime that have been used for personal identification. Before the trend of nanotechnology, various methods for latent fingerprint enhancement have been developed. However, these methods have a lack of sensitivity and selectivity to the fingerprint residues. In this case, they not only interact with fingerprint residues but

[*] Corresponding Author's E-mails: ritesh.shukla@ahduni.edu.in; shuklakritesh@gmail.com.

also with the background surface. In result, the contrast between the fingerprint and the background surface reduced and image become unidentifiable. To resolve this problem, nanoparticles (NPs) are then proposed for latent fingerprint development. Small size, high surface to volume ratio, unique optical and chemical properties of NPs make them appropriate for latent fingerprint development. Nanoparticles enhance the sensitivity and selectivity of the method used for fingerprint development in terms of its precise interaction with fingerprint residues and negligible interaction with background surface. It may lead to getting rid of the problem of background interference and best contrast image is formed. Several types of NPs including Gold, silver, metal oxides and quantum dots have been used for this purpose. These NPs have been applied as powder as suspension form for fingerprint development.

Precisely in this chapter, it evidently illustrated that nanotechnology has immense opportunities to nurture the field of latent fingerprint enhancement. However, it requires addressing some issues like method optimization, cost, efficiency and accuracy to conjugate with existing methods with the vision of eco-friendly in nature. In the future, the nanotechnology-based method may be implemented in real forensic casework, if these issues could be resolved.

Keywords: nanotechnology, forensic science, latent fingerprint

1. INTRODUCTION

Although Nanotechnology is a disputable technology, it is one of the most promising and emerging technology among the existed one [1] Nanotechnology has provided great benefit to the society and Forensic science in sense of various applications such as transforming the investigation process by making it more rapid, sensitive, accurate and easy to perform [2, 3]. These applications of nanotechnology make it indispensable and indisputable for forensics.

The advantage of nanotechnology is, it provides tremendous increased surface area to volume ratio in the nanomaterials as compared to the bulk material that helps to open a new avenue in surface science including latent fingerprint development [4]. The fingerprint is evidence that has been left behind at the scene of the crime by the criminals, unknowingly. The fingerprint is a unique pattern found at the tip of the fingers on the palmar surface of an individual's hand. Fingerprints left on the surface due to the action of secretion from the sweat and sebaceous gland of the palmar surface of the skin. Fingerprints can also explain the age, sex, diet, and medical condition of the individual [5]. The unique characteristics and the everlasting tendency of fingerprint make them ideal and important evidence for personal identification and individualization purpose. Since more than one-decade scientist and researchers have been actively involved in developing methods for latent fingerprints especially on porous surfaces (estimated 50") [6], which were remain undetected.

The basic mechanism of fingerprint development through powder method is based on generating a distinct contrast between the residue left by the finger and the underlying surface. But, the major drawback of the powder method of fingerprint development is, it also sticks on the underlying surface on which the fingerprint pattern present. It makes difficult to distinguish the contrast between the pattern and the surface. Hence, it is difficult to get a clear image of the pattern and so that identification is a challenging task [7]. To rectify this issue nanotechnology plays a vital role and introduces nanoparticles (NPs) (often in the form of nanopowders) that will adhere only to the necessary ridge area of the pattern and not the background surface, which naturally is an impediment. The unique properties of nanoparticles, which are its small size, flexibility to adjust with their surface properties make them suitable for the replacement [8]. The flexibility in surface modification plays a vital role that makes nanoparticles specific to precise targeting of components present in the fingerprint residues [9], which escalate the discernment of nanoparticles. Furthermore, distinctive optical properties of nanoparticles can enhance sensitivity via dropping the detection limit and it may solve the background interference for some extent. Nowadays, various types of nanopowder such as a metal oxide [10], gold [11], silver [12], silicon oxide [13] and semiconductors are (such as quantum dots) [14] being used for the latent fingerprint development on different surfaces [15]. Application of these NPs either in the form of dry powder or in the form of suspension (aqueous or organic solvent) used on latent fingerprints [16]. Powder method generally depends on physical process in which NPs (functionalized or non-functionalized) have an affinity with constituents (amino acid, protein, fatty compounds, organic and inorganic compounds) of fingerprint residues. The physical method of powder application may lead to low selectivity as well as triggering major health and safety issues. However, the application of the suspension method supports the physicochemical and chemical interactions with the residue [17]. In this order, Menzel in 2001 has successfully, applied photoluminescent CdS semiconductor nanocrystals capped with dioctyl sulfosuccinate to enhance the quality of fingerprint pattern [18]. Recently, Arshad et al. 2015 have synthesized the ZnO-SiO_2 nanopowder for fingerprint development [19]. The results exhibited that ZnO-SiO_2 nanopowder has excellent potential to envisage finger ridge detail at higher-level that shows superior discernibility than other commercially available white powders.

The precise mechanism comprises in the interaction between the nanoparticles and the constituents of fingerprint residues are still not fully elucidated. The scarcity of research work on the complexity of the fingerprint residues, play a major role in it [20-22]. Moreover, in situ studies of fingerprint residues are inadequate due to low quantity and thickness of materials and the total insubstantiality of the residues [22, 23].

Instead of fundamental principles involved in the detection process, research approaches focused on the application of NPs. This is another reason for unravel the mechanism of nanoparticles and fingerprint residues interaction. Technique optimization is a basic lacuna in nanoparticles based latent fingerprint development studies. This is the reason because of that; one study varies to other even same type of NPs used for fingerprint development. The results obtained so far with NPs are consequently not optimal [22, 23].

In this chapter, an attempt was made to integrate all the information related to nanoparticles, which have been used for latent fingerprint development.

1.1. Fingerprint

Fingerprints are the most common physical evidence found at the crime scene can be used for personal identification [24]. Generally, these fingerprints are three types including indented fingerprint, patent (visible) fingerprint, and latent (invisible) fingerprints. Common types of finger patterns are loop, arch, and whorl (Figure 1) [24]. When fingers touch on the soft surfaces such as putty, candle wax, or wet paint, three-dimensional fingerprints are formed, which known as an indented fingerprint. It visualized by naked eyes [24].

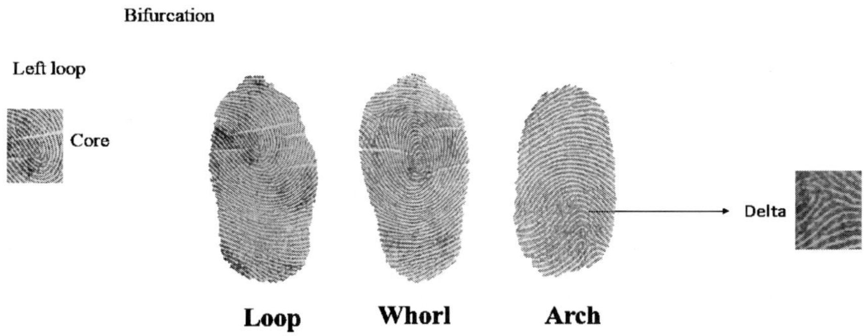

Figure 1. Types of finger patterns.

Visible fingerprints are straightforward to detect while latent prints require different developing/enhancement techniques for visualization [24, 25].

The complex mixture of natural secretors that comes from the secretory glands (eccrine, apocrine, and sebaceous) such as sweat and fatty compounds, as well as contaminants such as dirt, cosmetics or blood, are the main constituents of latent fingerprints and therefore, all visualization/enhancement methods rely on it [24]. The constituents of the latent fingerprint are mainly composed of water (99%) and minor amounts (up to 1%) of inorganic and organic compounds. The amount and secretory constituent of latent fingerprints may vary from individual to individual. This secretion may depend on some environmental conditions like temperature and humidity, exercise, stress, etc. that affect the production rate of eccrine sweat. This secretion also varies according to age, sex, medical condition and diet of an individual. Ultimately, these factors affect the quality of latent fingerprint after enhancement along with environmental conditions post-deposition [24].

Latent fingerprints are invisible prints so it requires some additional methods of enhancement through which print become visualize. The choice of latent fingerprint enhancement method depends on the composition of prints, type, and background of the substrate and on the ability of the technique to apply in sequence in the context of the case. Generally, optical, physical, or chemical enhancement methods are applying to visualize the latent fingerprints [24].

1.2. Latent Fingerprint Enhancement

Absorption, diffuse reflection, luminescence, UV absorption, and reflection are the optical methods used to visualize the latent fingerprints [26]. However, powder, small particle reagent and vacuum metal deposition are some physical methods and physical developer, multi-metal deposition, iodine, and cyanoacrylate are physical/chemical methods commonly used for latent fingerprint development [27]. In addition, ninhydrin and its analogs, metal complexation after ninhydrin treatment, DFO, 1,2-indanedione and genipin are some most prominent chemical methods used for latent fingerprint enhancement [28]. All of these methods have their own limitations therefore; latent fingerprint enhancement procedure is an open-ended research topic in the field of forensic research.

2. Nanotechnology in Forensics

The cutting edge of nanotechnology deals the exceptional opportunity to create functional materials devices and systems at a nanoscale (1-100 nm) level and holding the potential to revolutionize currently available technologies based on macro and micro scale [25, 29]. Since its multidimensional applications and advantages expand in all areas, it is referred to as universal pertinent technology that will have its impact on all areas from industries to society [8, 9, 30].

Surface area to volume ratio tremendously increases in various materials at their nano-scale as compared to its bulk counterparts [29, 31]. This physical property of nanoparticles opens new avenues in surface-based science including forensic sciences [32]. Some other physical phenomenon like decrease in size, crystal phase of the material, its doping properties, optical and electronic properties also play a vital role in this direction [33]. Nanotechnology associated with forensic via the development of nano-sensor, nano-based ink, nano-based security features in document and currency and nano-based methods (Fingerprint, drug, and explosive detection) for crime scene investigation [25, 29, 34].

2.1. Nanotechnology in Latent Fingerprint Development

Development of latent fingerprint at the crime scene is not an easy task. Although various physical and chemical methods have been applied, proper fingerprint analysis and identification will not be easy to achieve. In this order, many scientific groups have explored a variety of new methods to decipher latent fingerprint in order to assist the investigation process as well as enhance the efficiency of personal identification. Among all the physical and chemical methods, powder-dusting method is the most abundantly used for latent fingerprint enhancement. In the powder dusting method, the powder sticks to the fingerprint residue left by the finger and formed a visible characteristic pattern, which can be easily identifiable [31, 35-37]. During this process, most of the time conventional powders have shown their affinity not only with fingerprint residue but also with the substrate or surface on which the print was present. It leads to background noise, which conceals the fingerprint and making difficult to identify it. To overcome this issue Menzel et al. have introduced nano-powders to decipher the latent fingerprint more precisely as compared to conventional powders [36]. At present, various studies have revealed that rather than conventional powder, nano-powders provide better latent fingerprint development even on the tough surfaces like a bullet casing [38].

2.2. Nanopowder

To enhance the contrast, sensitivity, and selectivity, nano-powders have introduced as an efficient method for latent fingerprint development on non-porous surfaces [36]. Nano-powders are 1-100 nm size particulate matter that shows its efficiency to adhere with fingerprint residues only and provide fine ridge detail with greater contrast to the surface on which prints are present [25, 31, 32, 36]. At present, various types of nano-powders are being used for latent fingerprint development. Details of some prominently used nano-powders are cited herein in this chapter.

2.3. Metal Particles

Metallic nano-powders are reasonable highly stable, therefore used to retrieve latent fingerprint. The major advantage of using a powder method is ease in application and identification to retrieve the fingerprints. These nano-powders are much more efficient than their bulk counterparts. This technique is quite cost-effective too [25].

Among these metallic nano-powders, Gold nano-powders are the most commonly used for fingerprint detection. Properties like chemical inertness, high sensitivity, selectivity and resistance to oxidation make gold nano-powders a very strong component to bind with metallic surfaces [37]. Moreover, When a gold surface is immersed (or mixed, for nanoparticles) in a solution containing thiolated molecules, highly ordered self-assembled monolayers (SAMs) are rapidly formed [39]. These properties make Gold nano-powders, the best choice for latent fingerprint development. One of the studies, Choi et al. 2006 have reported that oleylamine stabilized Gold nano-powder interact with fatty compounds (lipophilic) present in fingerprint residues and deposited on it and finally, become visualize by naked eyes [40]. Moreover, Gold (III) has a tendency to reduce by Tin (II) and produce purple color while its size can be control by many factors like pH, temperature and agitation. These factors not only control its size but also produce in a variety of color. The reaction mechanism of gold nanopowder for fingerprint detection relied on redox reaction. In this reaction, Gold nano-powder act as a reducing agent to the fingerprint residues, which means gold particle lost its electron to the positively charged NaCl, an inorganic component of fingerprint residue, present into the fingerprints and generating précised visible print. In another study, to retrieve the latent fingerprints on non-bleached paper, low-density polyethylene, and polypropylene surfaces, Schnets has developed a modified method in which surfactant (Tween 20) reduces the non-selective deposition of gold nanoparticles on the background and its washing will also be effective that help to make better contrast [41]. With the help of this method, fingerprints can be visualized under ambient light as dark blue prints in good contrast background. This formulation may aid to retrieve the latent fingerprints within one-step only.

Furthermore, to promote and enhance the latent fingerprint development single and multi-metal deposition methods have developed.

2.4. Multi-Metal Deposition

Multi-metal deposition (MMD) is the two-tier chemical process in which small particle reagent (Gold) and of a physical developer (Silver) involved for latent fingerprint development. In the first step, a colloidal gold solution (pH 2.5-2.8) immersed with the latent fingerprint. Low pH plays a vital role in this step because it is a well-accepted theory that at acidic pH value (low pH), negatively charged gold nanoparticles (due to citrate ion) are attracted to positively charged residues present in fingerprint through electrostatic interaction [32, 42, 43]. In this first step, gold surface serves as a nucleation site for the growth of silver particle that will come in the second step of MMD [42, 44]. This step produces a poor contrast image due to the small size of colloidal gold. Therefore, to enhance the contrast of the print image with the deposition of a second metal such as Silver onto the gold surface is essentially required. Shown in (Figure 2). Thus, for the second step, gold treated print exposed with the silver physical developer that enhance the contrast image of the fingerprint

[45]. In this step, gold nanoparticles start aggregates as a growing site for the silver particles and fingerprint starts appearing as grey or black prints. Herein this method, in the presence of reducing agent, Silver ion obtained from the physical developer is reduced to the Silver metal on gold nucleation site.

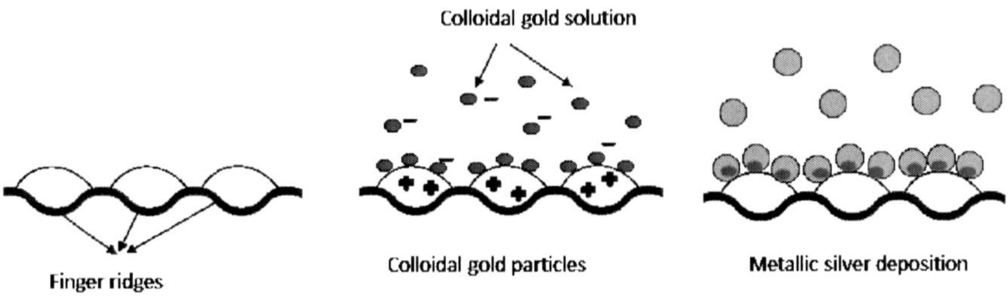

Figure 2. Schematization of the MMD (Multi-Metal Deposition) technique. Colloidal Gold particles are first deposited onto the fingermark residue and thereafter deposited gold particles are used as developing sites for the silver metal particles deposition.

Similar to the other methods, MMD also has its own limitations. The number of immersion step makes the procedure long and tedious. MMD method has applied by various research groups for latent fingerprint development in a different manner. Saunders first time introduced this method in 1989 for latent fingerprint development on various range of surfaces including porous/nonporous surfaces, wet surfaces, colored/non-colored surfaces, on adhesive tapes, and metal surfaces [41]. Subsequently, this method was used by Irrausch in 1991 and Allman et al. in 1992 [42, 46]. Irrausch demonstrated that MMD is the most effective method for latent fingerprint development on plastic, glass, white paper, tape, and polystyrene surfaces. However, this is the least effective method on the dry surface as compared to DFO or cyanoacrylate fuming method. In this order, Allman et al. 1992 have reported that MMD is least equal and often superior to cyanoacrylate fuming and VMD (vacuum metal deposition) methods for latent fingerprint development onto the plastics/polyethylene surfaces. Moreover, it has also been proved that MMD is the most promising method for fingerprint development on difficult surfaces like masking tape, beer bottle labels, and plastic gloves, cling film (Allman et al. 1992). In addition, Schnetz and Margot have modified MMD method using gold nanoparticles, alternative physical developer and accurate pH and found better result compared to the original method and named it as MMD II [12]. MMD II has shown highly effective and sensitive method on difficult semi-porous surfaces [12]. Furthermore, Beuce et al. 2007 have suggested another new MMD method in which gold nanoparticles modified with cyclodextrin molecules. Shown in (Figure 3). In this method, immersion procedures get reduced and gold nanoparticles functionalized with thiolated cyclodextrin in which dye molecules could be a trap [41].

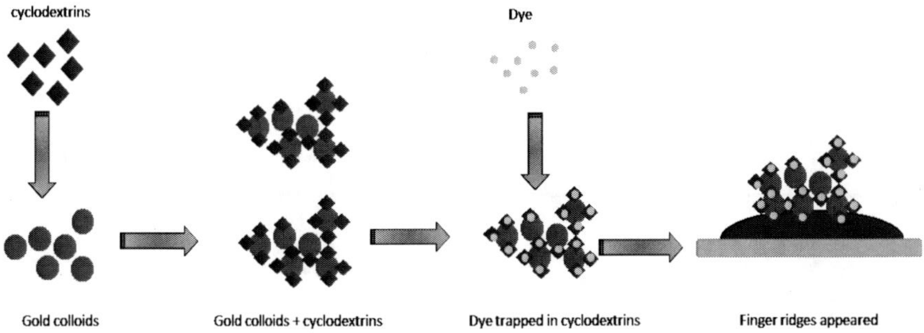

Figure 3. Experimental procedure for detection of latent fingerprint using modified colloidal gold particles. In the first step colloidal gold particles are functionalized with thiolated cyclodextrins. In the second step, a dye added in the functionalized solution and then latent fingerprint sample is finally immersed into it.

Recently, to overcome the limitation of MMD method, Stauffer et al. 2007 have proposed a new method known as single metal deposition (SMD). In this method, the second step (Silver developer step) of the MMD method replaced with gold enhancement step [47]. In this method, Gold colloids deposited on fingerprint serve as catalysts to precipitate the metallic gold (Au^0) from the hydroxy amine/gold chloride solution, thus increasing the size of the gold colloids. Shown in (Figure 4). It is based on the principle of electroless deposition reaction. The results obtained by SMD were very similar to those results, which obtained from MMD methods in terms of sensitivity and selectivity. In this method, Gold nanoparticles, synthesized using a reducing agent such as $NaBH_4$ in the presence of glucose as a stabilizing agent [48]. This method produces red color fingerprint after development.

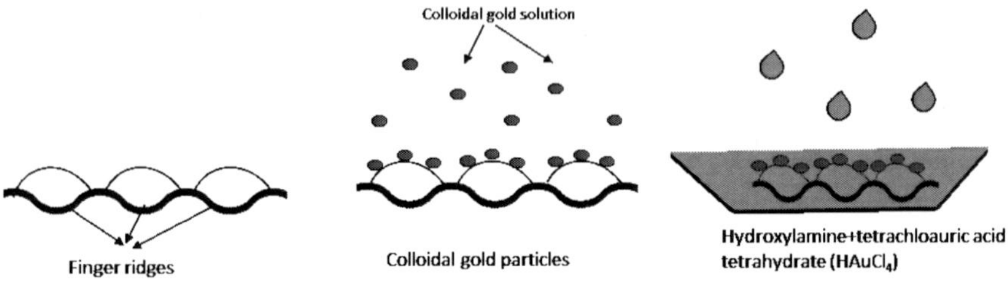

Figure 4. Schematic representation of SMD (Single Metal Deposition) technique.

2.5. Metal Oxide Nanoparticles

Several conventional metal oxide particles have been used to retrieve latent fingerprints from different surfaces.

2.5.1. Titanium Oxide Particles

Unique optical, electrical and photocatalytic properties of nano-sized TiO_2 make it the best choice for latent fingerprint development. Suspension of TiO_2 particles has used for latent fingerprint visualization on porous and non-porous surfaces [49]. It has reported that TiO_2 particles served as nucleation sites for the silver precipitation from the physical developer. In this study, TiO_2 suspension was prepared via acidification of white latex paint (44% TiO_2 and 56% Al_2O_5S) with aqueous citric acid at pH 3. However, in this study, it also observed that TiO_2 deposition after application of physical developer did not reveal any improvement in latent fingerprint over the conventional physical developer method. In addition, it has reported that when TiO_2 particle mixed with methanol it enhances the visibility of latent prints on non-porous and some semi-porous surfaces [50]. It was a two-step process in which first step TiO_2 mixed in a methanol carrier and spray on the print, thereafter, in second step print was rinsing with pure methanol. These encouraging results of latent fingerprint visualization on various surfaces through TiO_2 particles motivate to other research groups also to use this particle on some other surfaces. In this order, Schiemer et al. reported an interesting finding in which TiO_2 powder suspended in a surfactant solution was the best method for latent fingerprint visualization on the adhesive side of black electrical tape [51]. Similarly, Polimeni et al. 2004 and Cucè et al. 2004 have successfully developed latent fingerprints on wet surfaces using TiO_2 particle suspension (Polimeni et al. 2004, Cuce et al. 2004).

2.5.2. Zinc Oxide Particles

Zinc oxide is the second most abundantly used metal oxide nano-powder for latent fingerprint visualization. Zinc oxide can be used for both fresh and aged fingerprints on non-porous surfaces like glass, polyethylene and aluminum foil. These particles have fluorescence property that can be visualized under long UV light [52].

In another study, Leicester University scientists have developed a novel approach in which a nanoscale developer and an X-ray source together make it possible to retrieve the latent fingerprint on bullet casing even after rubbed or washed [52].

2.5.3. Iron Oxide Particles

Haque et al. in 1989 used black powder of iron oxide particles for latent fingerprint development on non-porous surfaces. The results of this study exhibited better sensitivity, clarity and contrast image of fingerprint as compared to conventional SPR(Small particle reagents) [53].

2.5.4. Silicon Dioxide Particles

SiO_2 NPs have handy abilities to modify its surface very easily for eg introducing a dye inside the SiO_2 NPs structure [13, 37, 42]. This surface modification property make SiO_2 NPs as a best choice to investigate the interactions taking place between fingerprint developing reagent and fingerprint residues.

2.5.5. Europium Oxide Particles

Generally, organic fluorescent compounds have a broad emission band and less excited state (lifetime 10-20 ns), however, Europium, a rare earth metal have tendency opposite to earlier one that is narrow emission band and a long excited-state (lifetime 50-100 μs) which make this rare earth metal ideal for detection/visualization of invisible items. Keeping in view, Menzel 1990, has introduced Europium for latent fingerprint development. In another study, Menzel et al. 2005 also introduced amine-functionalized europium oxide nanoparticles for latent fingerprint development [54]. In this study, functionalized nanoparticles targeted the carboxylic acid functionalities present in fingerprint residues. The result of this study exhibited the better contrast image of the developed fingerprint because the background fluorescence of the surface was significantly reduced. Shown in (Figure 5).

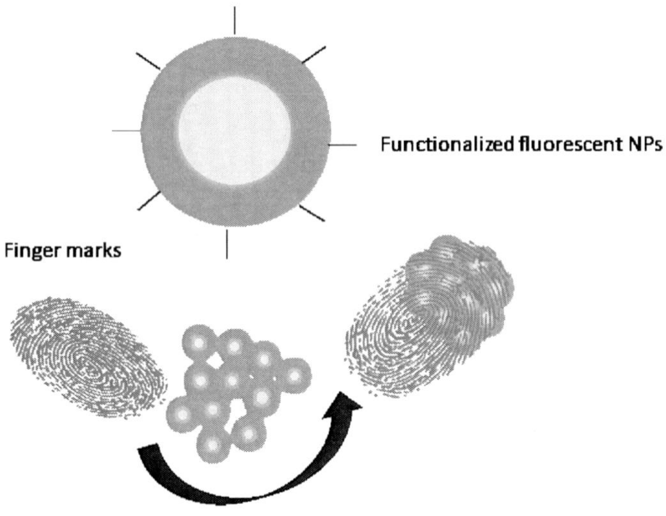

Figure 5. Schematic approach of fingerprint detection using functionalized fluorescent nanoparticles.

2.6. Metal Sulfide Particles

Similar to metal and metal oxide particles, metal sulfide particles have also used for latent fingerprint visualization. Molybdenum disulfide and Cadmium sulfide are the most abundantly used for this purpose.

2.6.1. Molybdenum Disulfide

Preparation of Molybdenum disulfide (MoS_2) suspension has resembled with small particles reagent* (suspension of fine particles in detergent solution) whereupon the application MoS_2 particle reacts with the fatty component of the latent fingerprint residues and becomes visualized as a grey print image.

This SPR based MoS_2 method applied on various surfaces including porous or non-porous, adhesive surfaces and successfully develop prints on these surfaces. This method most efficiently worked on wet or dusty surfaces [6, 55]. SPR method of latent fingerprint development has generally applied by immersion or spraying method. The immersion method is comparatively highly sensitive and more effective on fresh prints than aged ones. The spray method is considerably less effective [49].

*SPR Method - SPR method for latent fingerprint development was first introduced in the 20th century [45]. Subsequently, Goode and Morris have described a detailed procedure and formulation of SPR in 1983 [25]. This method relies on the interaction of fine particles with the oily or fatty component of fingerprint residue [24, 56]. Suspension of small particle reagent is prepared by addition of fine particles, surfactant, and water. In this method, regular powders and fluorescent powders are most commonly used. Generally, in this method, sodium dodecyl sulfate (SDS) is used as a surfactant [57]. Here, the role of surfactant is to provide affinity to bind small particles with fatty components of the fingerprint residues [57]. Surfactant also provides relative stability to these components (small particles and fatty residues), which make this method appropriate for aged fingerprints development. Additionally, this method is also useful for developing fingerprints on wetted non-porous surfaces. In this method, the concentration of the surfactant, play a vital role as its lower concentration produce poor results while higher concentration produces unwanted/undeposited particles to be wash away by water. In this order, Frank and Almog [58] recommended a new white SPR formulation that contains zinc carbonate powder, which aid to develop latent fingerprints for dark surfaces. In addition, Springer and Bergman have developed a fluorescent SPR formulation in which an ethanolic solution of the Basic Yellow 40 dye (BY40) added into the SPR stock solution [59]. In this study, fingerprints treated with BY40 SPR visualized at Excitation 450 nm/ emission at 550 nm on dark/multicolor surfaces. Additionally, Zamir et al. in

2002 recommended that the SPR method also aid to develop bloody fingerprints that could not interfere for consequent DNA analysis [60].

SPR method is an efficient add-on to the powder method for developing prints on wet, immersed and aged fingerprints on non-porous surfaces [57].

2.6.2. Cadmium Sulfide Nanocrystals

Cadmium sulfide (CdS) method more effectively worked on the surfaces, which are previously treated, with cyanoacrylate ester fuming method for latent fingerprint development. CdS nanocrystals capped with dioctyl sulfosuccinate in heptane or a mixture of hexanes that help to visualize the fingerprints on a soft drink cane and aluminum foil, which is prior, treated with cyanoacrylate ester fuming method [61]. In this study, it also observed that if fingerprints were not previously treated, with cyanoacrylate ester fuming method then prints were not developed because heptane and hexane have a tendency to obliterate ridge details.

In addition, Menzel et al. have developed fluorescent CdS-dendrimer nanocomposites which were branched tree-like structure in which amine group attached on terminal end, successfully react with the carboxylic acid group present in fingerprint resides and visualized fingerprints [62]. In this study, CdS-dendrimer nanocomposites were prepared by adding solutions of cadmium nitrate and sodium sulfide to a dendrimer solution. This nanocomposite produces blue-green luminescence. This method visualized prints on different surfaces including aluminum foil, polyethylene (fumed) and paper. In another study, CdS nanocomposites with a carboxylate terminal-functionalized dendrimer were studied [63]. In this study, it was reported, that carboxylate functional groups which were present in nanocomposite can react with amino acids or proteins found in fingerprint residue. In this method, developed prints (orange color) were faint without heating.

In the aforementioned studies, it has been clearly exhibited that nano-based methods are better than their conventional methods of latent fingerprint development. Brief descriptions of these nano-based methods for latent fingerprint development are schematically shown in (Figure 2-5). Moreover, some significant studies of latent fingerprint development using nano-based methods are shown in Table 1.

Table 1. Finger print development using nanoparticles based method

Methods	Surface	Role	References
MMD (multi-metal deposition)	Porous and non-porous both	Colloidal gold nanoparticles as a first metal and metallic silver particles as a second reducing metal.	(Saunders 1989)
SMD (single metal deposition)	Porous, non-porous both and surfaces like polystyrene, thermal paper, wet surface.	Colloidal gold nanoparticles as a first metal and hydroxylamine + tetrachloauric acid tetrahydrate	(Schnetz and Margot 2001)
		Metal particles	
Gold NPs	Silicon wafers and paper Strips	Au-NPs stabilized by n-alkanethiol chains and suspended in petroleum ether, stick to fingerprints ridges and improve the development of latent fingerprints by the Ag-PD method.	(Sametband et al. 2007)
	Non-bleached paper, low-density polyethylene, and polypropylene surfaces	Au-NPs functionalized with thiolated cyclodextrins and ridges appear in the form of dark-blue prints.	(Becue et al. 2007)
	Non-porous surfaces	Zinc oxide deposited on gold nanoparticles via in situ approach and obtained luminescent fingerprints on black surfaces	(Becue et al. 2008)
	Paper Strips	Purple colour fingerprints was observed on paper strips in single step procedure via In situ growth of AuNPs.	(Hussain et al. 2010)
	Non-porous surfaces	Aged and dried latent fingerprints were successfully enhanced on non-porous surface by using the AuNPs conjugated with anti-L-amino acid antibodies.	(Spindler et al. 2011)
	Porous surface	AuNPs capped with mercapto carboxylic acids followed by silver precipitation step, develop a negative impression of latent fingerprints on paper surface.	(Shenawi et al. 2013)
		Poly(styrene-alt-maleic anhydride)-b-polystyrene (PSMA-b-PS) functionalized with AuNPs develop colorimetric as well as photoacoustic imaging of latent fingerprint.	(Song et al. 2015)

Table 1. (Continued)

Methods	Surface	Role	References
	Non-porous surfaces	Lysozyme- binding aptamer-conjugated Au nanoparticles (NPs) has used to visualize the latent fingerprint.	(Su et al. 2016)
Metal oxide particles			
TiO_2 NPs	Non porous surface	TiO_2 particles are used as white powder or white reagent which helps to develop prints on dark surfaces. It also works on wet medium	(Wade 2002)
	Blood and semi-porous surfaces	TiO_2 particles with methanol enhance the visualization of finger prints in blood and semi porous surfaces.	(Bergeron 2003)
	Wet surface	TiO_2 particles generate fingerprints on wet surfaces.	(Polimeni et al. 2004, Cuce et al. 2004)
		TiO_2 can be used as paste with brush or spray.	(Williams and Elliott 2005)
	Perylene dianhydride, adsorbed onto TiO_2 nanoparticles	Strong fluorescence at 650–700 nm when excited at 505 nm	(Choi et al. 2007)
Zinc oxide particles (ZnO)		ZnO particles are used as fluorescent pigment for detection.	(Choi et al. 2008a)
	Prepared by the in situ deposition of zinc oxide onto the gold nanoparticles.	visible luminescence at approximately 580 nm when excited with UV light	(Becue et al. 2008)
	Non porous surface	developing prints on dry and wet surfaces excellent third-level ridge details with minimal background staining	(Arshad et al. 2015)
Iron oxide particles		Black powder of particles show better sensitivity, clarity and contrast for finger marks.	(Haque et al. 1989)
			(Choi et al. 2006)

Methods	Surface	Role	References
Silica nanoparticles	Doped with rhodamine B and 6 G	Fluorescent image formed when excited with UV light (415 nm)	(Theaker et al. 2008)
	Modified by 4-(Chloromethyl) phenyltrichlorosilane		(Huang et al. 2015)
	Fluorescent conjugated oligomer	Highly efficient blue emission	(Zhang et al. 2017b)
	Loaded with Methylene blue (MB) molecules		(Zhang et al. 2017a)
		Latent blood fingermarks on black plastic bags, Print visualized with green emission	(Meng et al. 2018)
Europium oxide particles		Particles focus on carboxylic acid constitutes of finger marks and can be seen fluorescently.	(Menzel et al. 2005)
Metal sulfide particles			
Molybdenum disulfide (MoS_2)		These particles focus on fatty acid constitutes of fingers. Spray method is more effective in this.	
Cadmium sulfide nanocrystals		CdS nanocomposites are used for fumed and un-fumed finger marks.	(Moret et al. 2013)
Reagents(chemical)			
Silver nitrate ($AgNO_3$)	Porous and untreated wood	Mostly used to detect chloride ions contained in papillary secretions.	(Aubert 1878, Olsen 1978)
Amino acid reagents	Paper fibres	Used to detect sweaty hands and finger marks.	
1. Ninhydrin (2,2-dihydroxy-1,3-indanedione)	Porous surfaces like paper and cardboards.	Can detect primary and secondary amines contained in amino acids or in polypeptides and show purple marks.	(Odén and VON HOFSTEN 1954)

Table 1. (Continued)

Methods	Surface	Role	References
2. DFO (1,8-diazafl uoren-9-one) 3. 1,2-indanedione	On porous surfaces	It gives faint pink colour to finger ridges but extremely luminescent Show pink colour marks at room temperature with luminescence. It can be alternative for DFO.	(Wilkinson 2000) (Wilkinson 2000, Wiesner et al. 2001), (Kasper et al. 2002)
p-4-Dimethylaminocinnamaldehyde (pDMAC)	Porous surfaces.	Sweat contains urea and that can be detected by this chemical. Show red and luminescent marks.	(Sasson and Almog 1978, Brennan et al. 1995)
Osmium tetroxide (OsO_4) and ruthenium tetroxide (RuO_4)		These two chemicals target double bonds of the unsaturated organic compounds found in a fingermark deposit through an oxidation mechanism	(Grzegorzewska and Filbrandt 2004)
Physical detection methods			
Powder dusting	Technique is limited to smooth and non-porous surfaces	Applied through brush and it adheres to sticky, moistly and greasy surface.	(Reiss 1911, Lee and Gaensslen 2001, Robertson et al. 2004)
Camphor	Firearms cartridges	Camphor is burnt near to finger marks and allows the black soot to settle down on marks, finger ridges show black linings.	(Waldoch 1993)
Iodine fuming	Porous and non- porous surfaces both.	The vapor absorbs lipid from the finger marks and form dark black image.	(Mashiko and Ishizaki 1977, Haque et al. 1983)

Table 2. Finger print development using Quantum dot based method

	Surface	**Procedure**	**References**
QDS(Powdering method)			
Cds/chitosan nanocomposits (NCs)	Aluminium substrate	The latent fingerprints were fumed with cyanoacrylate ester and then NCs were applied	(Dilag et al. 2009)
Cds/PPH (porous phosphate heterostructure NCs	Plastic, glass, steel, ceramic, and wood surfaces	PPH are hybridized with mercaptopropyl group and used directly on latent finger marks. It observed under excitation of 450nm.	(Algarra et al. 2011)
CdTe- SiO_2 (silicon dioxide) nanomaterials (NMs)	glass, polymer, plastic, rubber, marble, aluminum, and paper	SiO_2 coated CdTe makes a core shell structure which shows great sensitivity and enhanced fluorescence.	(Gao et al. 2011a)
CdTe/MMT (montmorillonite)NCs	Painted wood, leather, polymer and glass	NCs deposits on latent finger marks, especially on papillary ridges and observed under 365nm.	(Gao et al. 2011b)
QDS(Liquid method)			
CdS-DSS (dioctyl sodium sulfosuccinate)	metallic and sticky substrates	CdS-DSS mixture were dissolved in heptane or hexane and perfumed the surface with cyanoacrylate ester and observe fingermarks. The dried sample can be observed under UV light	(Menzel et al. 2000a)

Table 2. (Continued)

	Surface	Procedure	References
CdS/PAMAM (polyamidoamine dendrimers) NCs	Aluminum foil	Marks are first perfumed with cyanoacrylate ester and then immerced in to the CdS/PAMAM Mixture for overnight at room temperature.	(Jin et al. 2008)
CdSe QDs	stabilized with n-alkaneamines in organic solution	Direct visualization of latent fingerprints on wet nonporous surfaces such as silicon.	(Sametband et al. 2007)
	stabilized with mercaptoacetic acid	fluorescence when excited by 365 nm UV light	(Pan et al. 2000) (Wang et al. 2009)
CdSe with thioglycolic acid	Functionalized with thiolated porous phosphate heterostructures	Unfumed finger marks are immersed in the solution and dried and observed under UV light of 380nm	(Wang et al. 2009)
		Fluorescent image showed a red emission at 576 nm combined with other at 473 nm	(Algarra et al. 2013)
		Satisfactory performance on sticky side of light-colored tapes; while, the latter worked significantly better on dark-colored tapes	(Wang et al. 2014)
CdTe QDs		Finger ridges are observed	(Liu et al. 2010)

	Surface	**Procedure**	**References**
1. Positively charged CdTe QDS with hydrazine hydrate		Enhanced affinity and effectivity, compared with the use of negatively charged CdTe QDs. Can excited under 365nm.	(Gao et al. 2011b)
2. Negatively charged CdTe QDS with TGA		Lower affinity and effectivity but excitation at 360nm	(Gao et al. 2011b)
QDS (blood finger marks)			
CdTe QDS with TGA modified	Nonporous substrates, including transparent polypropylene, black polypropylene, glass, and aluminum foil	Blood finger marks are first immersed in 5-sulfosalicylic acid and then immersed in this solution.	(Becue et al. 2009)
CdTe QDs dopped with copper	Detection of blood fingerprint on non-porous surfaces	Red-shift in emission spectra	(Moret et al. 2013)
CdTe QDs capped with mercaptosuccinic acid (MSA)	adhesive tape, transparent tape, aluminum foil and stainless steel		(Cai et al. 2013)
Mn doped ZnS QDs	glass, aluminium foil and polymethylpentene plastic,		(Cheng et al. 2008)
3-mercaptopropionic acid (MPA)-modified ZnS:Cu (copper doped zinc sulphide) QDs NPs	Nonporous substrates, including transparent polypropylene, black polypropylene, glass, and aluminum foil	It is similar to Becue,2009 et al. Finger marks are visible under 300-400nm.	(Moret et al. 2013)
Copper-doped	blue-green luminescence emission		(Moret et al. 2013)
Mn-Doped ZnS QDs	Nonporous surface react with eccrine	Orange fluorescence emission	(Xu et al. 2014)

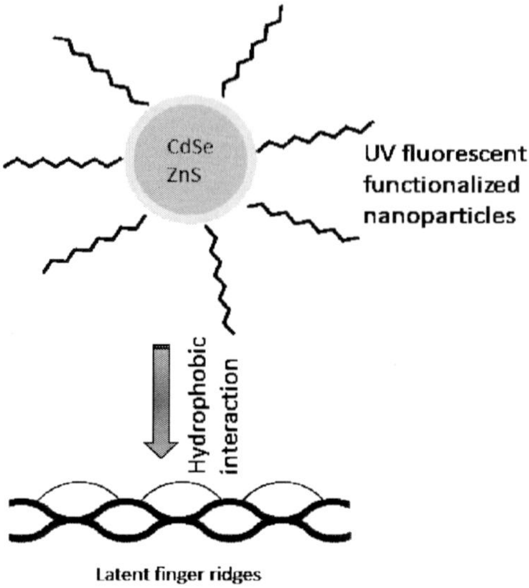

Figure 6. Schematic illustration of latent fingerprint development using functionalized quantum dots.

2.7. Quantum Dots

Due to unique optical and chemical properties, quantum dots (QDs) are the new emerging agent for retrieval of latent fingerprints on different surfaces. QDs have a tendency to produce strong visible light at blue or UV range of wavelength [77, 78]. QDs exhibit various optical properties including high intensity, good chemical and photostability, superficial surface modification, broad absorption spectra, size- and composition-tunable emission, narrow emission spectra, high quantum yield, and low toxicity [79, 80]. In addition, an exclusive versatile chemical property of QDs to modify the surface with a range of functional ligands, open new promises for latent fingerprint development with extraordinary contrast, sensitivity and selectivity [81, 82]. Hitherto different types of QDs including CdS, CdSe, and CdTe QDs have been used for retrieving latent fingerprint [83].

2.7.1. QDs in Powder Method

Similar to the conventional powder method, QDs are also permitted to adhere to the aqueous or oily components of the fingerprint residues, during latent fingerprint development. The interaction between QDs and fingerprint residues are predominantly governed by physical adsorption and electrostatic attraction phenomenon. When QDs exposed to air after interacting with fingerprint residues, due to the high oxidation state of QDs, fluorescent intensity emission get reduced which leads to poor developing contrast. It shows that longtime fingerprint conservation becomes problematic. To overcome this issue, QDs are conjugated with specific materials that form QDs-nanocomposites (QDs-NCs) [57].

2.7.2. QDs in Liquid Method

When QDs used in the liquid method for latent fingerprint development, they interact with the inorganic and organic component of the fingerprint residues. This interaction is based on physical adsorption and electrostatic attraction phenomenon. To achieve the adsorption interaction, QDs' surface was frequently modified by several explicit functional ligands.

QDs are also used for blood fingerprint development. In this method, QDs specifically interact with hemoglobin, a major blood component. To develop blood fingerprints, QDs' surface modified by specific functional ligands that have a higher affinity to conjugate with hemoglobin [57].

In various studies, latent fingerprints have developed by functionalized or non-functionalized QDs, descriptive details are shown in Table 2. A schematic representation is shown in (Figure 5).

CONCLUSION

In conclusion, we elucidated that there are various methods available for latent fingerprint visualization. The available methods could be targeting only one or two subsets of the fingerprint residue however, the complexity of the residue not fully explored. This reason clarifies why the claim for better reagent for latent fingerprint enhancement continued over the years. Although methods are developing with rapid pace method optimization is lagging behind. In this context, nanotechnology has proved to be a promising area of research.

Generally, it is reflected that nanotechnology aid in developing new materials and reagents having superior physicochemical characteristics compared to their bulk counterparts. It offers new research prospects in surface sciences. Since latent fingerprint development based on the surface phenomenon, it is evident that nanotechnology is a leading entrant to change the output of the results through a completely different approach. This chapter appraised the status of nanotechnology-based methods applied to latent fingerprint development. Precisely, this chapter emphasized on surface-specific various types of nanoparticles and their interaction with fingerprint residues. Some methods are still in a preliminary stage and need more work on it before implementation in casework. More consideration and research are urgently needed for future advancement of nanotechnology-based latent fingerprint development methods.

FUTURE PERSPECTIVES

This chapter evidently illustrated that nanotechnology has immense opportunities to nurture in the field of latent fingerprint enhancement. However, its routine application in the field of forensic casework is still far away and requires being address on some important issues like methods optimization, cost, efficiency, accuracy to conjugate with existing methods and prospects to safe human and environment in near future. Nanotechnology's Applications in the field of forensic practice is still inadequate. Scarcity of method optimization and multiple influencing physicochemical properties of nanoparticles are the major factors, which create obstacles to bring nanotechnology-based methods to apply in the real world of forensic casework. Hence, foremost futuristic research in the field of forensic and latent fingerprint enhancement should be focused on method optimization and stability of nanoparticles' properties that aid to enhance the strength and consistency of nanotechnology for its application in the real forensic casework. Indeed, this technology would transform the forensic investigation process and reduce the number of pending and unsolved criminal cases. In the future, the application of nanotechnology in forensic/latent fingerprint enhancement would shift the paradigm of forensic research and open new avenues, which effectively translate the laboratory to the real world.

REFERENCES

[1] Teeuw WB, Vedder AH, Custers BH, Dorbeck-Jung BR, Faber EC, Iacob SM, et al. *Security Applications for Converging Technologies:* Boom Juridische uitgevers; 2008.

[2] Lodha A, Pandya A, Shukla R. Nanotechnology: An Applied and Robust Approach for Forensic Investigation. *Forensic Res Criminol Int J.* 2016;2(1):00044.

[3] Gooch J, Daniel B, Abbate V, Frascione N. Taggant materials in forensic science: A review. *TrAC Trends in Analytical Chemistry.* 2016;83:49-54.

[4] Shukla RK. Occupational Exposure of Nanoparticles in Forensic Science: A Need of Safe Use. *Int J Forensic Sci Pathol.* 2013;1(3):7-10.

[5] Penn DJ, Oberzaucher E, Grammer K, Fischer G, Soini HA, Wiesler D, et al. Individual and gender fingerprints in human body odour. *Journal of the Royal society interface.* 2006;4(13):331-40.

[6] Lee HC, Gaensslen R. Methods of latent fingerprint development. *Advances in fingerprint technology.* 2001;2(105-176):10.

[7] Pandya A, Shukla RK. New perspective of nanotechnology: role in preventive forensic. *Egyptian Journal of Forensic Sciences.* 2018;8(1):57.

[8] Jones N, Mansour D, Stoilovic M, Lennard C, Roux C. The influence of polymer type, print donor and age on the quality of fingerprints developed on plastic substrates using vacuum metal deposition. *Forensic science international.* 2001;124(2-3):167-77.

[9] Jones N, Kelly M, Stoilovic M, Lennard C, Roux C. The development of latent fingerprints on polymer banknotes. *Journal of Forensic Identification.* 2003;53(1):50.

[10] Pfister R. *The optical revelation of latent fingerprints.* 1985:64- 70.

[11] Cantu A. Silver physical developers for the visualization of latent prints on paper. *Forensic Science Review.* 2001;13(1):29-64.

[12] Schnetz B, Margot P. latent fingermarks, colloidal gold and multimetal deposition (MMD): Optimisation of the method. *Forensic Science International.* 2001;118(1):21-8.

[13] Lennard CJ, Margot PA, Sterns M, Warrener RN. Photoluminescent enhancement of ninhydrin developed fingerprints by metal complexation: structural studies of complexes formed between Ruhemann's purple and group IIb metal salts. *Journal of Forensic Science.* 1987;32(3):597-605.

[14] Pounds A, Grigg R, Mongkolaussavaratana T. The use of 1, 8-diazafluoren-9-one (DFO) for the fluorescent detection of latent fingerprints on paper. A preliminary evaluation. *Journal of Forensic Science.* 1990;35(1):169-75.

[15] Torgal FP, Jalali S. Toxicity of Construction and Building Materials. *Eco-efficient Construction and Building Materials:* Springer; 2011. p. 19-33.

[16] Chen H, Ma R-l, Chen Y, Fan L-J. Fluorescence development of latent fingerprint with conjugated polymer nanoparticles in aqueous colloidal solution. *ACS applied materials & interfaces.* 2017; 9(5):4908-15.

[17] Wallace-Kunkel C, Lennard C, Stoilovic M, Roux C. Optimisation and evaluation of 1, 2-indanedione for use as a fingermark reagent and its application to real samples. *Forensic science international.* 2007;168(1):14-26.

[18] Menzel ER. Recent advances in photoluminescence detection of fingerprints. *The Scientific World Journal.* 2001;1:498-509.

[19] Arshad A, Farrukh MA, Ali S, Khaleeq-ur-Rahman M, Tahir MA. Development of Latent Fingermarks on Various Surfaces Using ZnO-SiO2 Nanopowder. *Journal of forensic sciences.* 2015; 60(5):1182-7.

[20] Wiesner S, Springer E, Sasson Y, Almog J. Chemical development of latent fingerprints: 1, 2-indanedione has come of age. *Journal of Forensic Science.* 2001;46(5):1082-4.

[21] Almog J, Cohen Y, Azoury M, Hahn T-R. Genipin—a novel fingerprint reagent with colorimetric and fluorogenic activity. *Journal of Forensic Science.* 2004;49(2):1-3.

[22] Becue A, Moret S, Champod C, Margot P. Use of quantum dots in aqueous solution to detect blood fingermarks on non-porous surfaces. *Forensic Science International.* 2009;191(1-3):36-41.

[23] Yang RQ, Wang YG, Xia BB, Wang YJ, Liu JJ, editors. Application of CdTe quantum dots to development fingerprints on adhesive surfaces. *Materials Science Forum;* 2011: Trans Tech Publ.

[24] Saferstein R, Graf SL. Evaluation of a reflected ultraviolet imaging system for fingerprint detection. *Journal of Forensic Identification.* 2001;51(4):385.

[25] Goode G, Morris J. *Latent fingerprints: a review of their origin, composition and methods for detection:* Atomic Weapons Research Establishment; 1983.

[26] Lin S-S, Yemelyanov KM, Pugh Jr EN, Engheta N. Polarization-based and specular-reflection-based noncontact latent fingerprint imaging and lifting. *JOSA A.* 2006;23(9):2137-53.

[27] Yamashita B, French M, Bleay S, Cantu A, Inlow V, Ramotowski R, et al. *The fingerprint sourcebook.* National Institute of Justice/NJRS, Rockville. 2010:1-67á.

[28] Jelly R, Patton EL, Lennard C, Lewis SW, Lim KF. The detection of latent fingermarks on porous surfaces using amino acid sensitive reagents: a review. *Analytica Chimica Acta.* 2009;652(1-2):128-42.

[29] Pfister R, editor. The optical revelation of latent fingerprints. *Journal of the Forensic Science Society;* 1984: Forensic Science SOC Clark House, 18A Mount Parade, Harrogate, N Yorkshire ....

[30] Jones N, Stoilovic M, Lennard C, Roux C. Vacuum metal deposition: factors affecting normal and reverse development of latent fingerprints on polyethylene substrates. *Forensic science international.* 2001;115(1-2):73-88.

[31] Burow D, Seifert D, Cantu AA. Modifications to the silver physical developer. *Journal of forensic sciences.* 2003;48(5):1094-100.

[32] Baschong W, Stierhof YD. Preparation, use, and enlargement of ultrasmall gold particles in immunoelectron microscopy. *Microscopy research and technique.* 1998;42(1):66-79.

[33] Hodes G. When small is different: some recent advances in concepts and applications of nanoscale phenomena. *Advanced Materials.* 2007;19(5):639-55.

[34] Margot P, Lennard C. Fingerprint detection techniques: *Université de Lausanne Institut de police scientifique et de criminologie;* 1994.

[35] Sodhi GS, Kaur J. Powder method for detecting latent fingerprints: a review. *Forensic science international.* 2001;120(3):172-6.

[36] Ramotowski R, Cantu A, Joullié M, Petrovskaia O. 1, 2-Indanediones: a preliminary evaluation of a new class of amino acid visualizing compounds. *Fingerprint Whorld.* 1997;23(90):131-40.

[37] Jones N. *Metal deposition techniques for the detection and enhancement of latent fingerprints on semi-porous surfaces.* 2002.

[38] Babin B. Development of Friction Ridges on a 9mm Bullet Casing. *Identification Canada.* 2010;33:103-6.

[39] Dubois LH, Nuzzo RG. Synthesis, structure, and properties of model organic surfaces. *Annual review of physical chemistry.* 1992;43(1):437-63.

[40] Choi MJ, McDonagh AM, Maynard PJ, Wuhrer R. Preparation and evaluation of metal nanopowders for the detection of fingermarks on nonporous surfaces. *Journal of Forensic Identification.* 2006;56 (5):756.

[41] Becue A, Champod C, Margot P. Use of gold nanoparticles as molecular intermediates for the detection of fingermarks. *Forensic science international.* 2007;168(2-3):169-76.

[42] Allman D, Maggs S, Pounds C. The use of colloidal gold/multimetal deposition for the detection of latent fingerprints–a preliminary evaluation. Central Research and Support Establishment, *Home Office Forensic Science Service*, Aldermaston, UK, Report. 1992;747:1ą24.

[43] Choi MJ, McDonagh AM, Maynard P, Roux C. Metal-containing nanoparticles and nano-structured particles in fingermark detection. *Forensic Science International.* 2008;179(2-3):87-97.

[44] Festag G, Steinbrück A, Csáki A, Möller R, Fritzsche W. Single particle studies of the autocatalytic metal deposition onto surface-bound gold nanoparticles reveal a linear growth. *Nanotechnology.* 2006;18(1):015502.

[45] Gaensslen RE, Ramotowski R, Lee HC. *Advances in fingerprint technology:* CRC press; 2001.

[46] Irrausch F. L'utilisation de l'or colloidal dans la révelation des empreintes digitales: la déposition multimetallique [The use of colloidal gold in the revealing of fingerprints: the multimetallic deposition]. *Seminaire 4eme annee Institut de police scientifique et de criminologie,* Université de Lausanne. 1991.

[47] Stauffer E, Becue A, Singh KV, Thampi KR, Champod C, Margot P. Single-metal deposition (SMD) as a latent fingermark enhancement technique: an alternative to multimetal deposition (MMD). *Forensic Science International.* 2007;168(1):e5-e9.

[48] Gao D, Li F, Song J, Xu X, Zhang Q, Niu L. One step to detect the latent fingermarks with gold nanoparticles. *Talanta.* 2009;80(2):479-83.

[49] Cantu A. *Notes on some latent fingerprint visualization techniques developed by Dr. George Sanders,* US Secret Service, Forensic Service Division, Washington, DC. 1996.

[50] Bergeron J. Development of bloody prints on dark surfaces with titanium dioxide and methanol. *Journal of forensic Identification.* 2003;53(2):149.

[51] Schiemer C, Lennard C, Maynard P, Roux C. Evaluation of techniques for the detection and enhancement of latent fingermarks on black electrical tape. *Journal of Forensic Identification.* 2005;55(2):214.

[52] Choi MJ, McBean KE, Ng PHR, McDonagh AM, Maynard PJ, Lennard C, et al. An evaluation of nanostructured zinc oxide as a

fluorescent powder for fingerprint detection. *Journal of Materials Science*. 2008;43(2):732-7.

[53] Haque F, Westland AD, Milligan J, Kerr FM. A small particle (iron oxide) suspension for detection of latent fingerprints on smooth surfaces. *Forensic Science International*. 1989;41(1-2):73-82.

[54] Menzel ER, Schwierking JR, Menzel LW. Functionalized europium oxide nanoparticles for fingerprint detection: a preliminary study. *Journal of Forensic Identification*. 2005;55(2):189.

[55] Lennard CJ, Champod C, Margot P, Stoilovic M. *Fingerprints and other ridge skin impressions:* CRC press; 2016.

[56] Ramotowski R. *Lee and Gaensslen's advances in fingerprint technology:* CRC Press; 2012.

[57] Wang M, Li M, Yu A, Zhu Y, Yang M, Mao C. Fluorescent nanomaterials for the development of latent fingerprints in forensic sciences. *Advanced functional materials*. 2017;27(14):1606243.

[58] Frank A, Almog J. Modified SPR (Small Particle Reagent) for Latent Fingerprint Development on Wet, Dark Objects. *Journal of Forensic Identification*. 1993;43(3):240-4.

[59] Springer E, Bergman P. A fluorescent small particle reagent (SPR). *Journal of Forensic Identification*. 1995;45(2):164-8.

[60] Zamir A, Oz C, Leifer A, Geller B. The effect of small particle regent employed as a fingerprint enhancement technique on subsequent STR typing from bloodstains. *Journal of Forensic Identification*. 2002;52(6):691.

[61] Menzel ER, Savoy SM, Ulvick SJ, Cheng KH, Murdock RH, Sudduth MR. Photoluminescent semiconductor nanocrystals for fingerprint detection. *Journal of Forensic Science*. 2000;45(3):545-51.

[62] Menzel E, Takatsu M, Murdock R, Bouldin K, Cheng K. Photoluminescent CdS/dendrimer nanocomposites for fingerprint detection. *Journal of Forensic Science*. 2000;45(4):770-3.

[63] Bouldin KK, Menzel ER, Takatsu M, Murdock R. Diimide-enhanced fingerprint detection with photoluminescent CdS/dendrimer nanocomposites. *Journal of Forensic Science*. 2000;45(6):1239-42.

[64] Sametband M, Shweky I, Banin U, Mandler D, Almog J. Application of nanoparticles for the enhancement of latent fingerprints. *Chemical Communications.* 2007(11):1142-4.

[65] Becue A, Scoundrianos A, Champod C, Margot P. Fingermark detection based on the in situ growth of luminescent nanoparticles—Towards a new generation of multimetal deposition. *Forensic Science International.* 2008;179(1):39-43.

[66] Hussain I, Hussain SZ, Ihsan A, Rehman A, Khalid ZM, Brust M, et al. In situ growth of gold nanoparticles on latent fingerprints—from forensic applications to inkjet printed nanoparticle patterns. *Nanoscale.* 2010;2(12):2575-8.

[67] Spindler X, Hofstetter O, McDonagh AM, Roux C, Lennard C. Enhancement of latent fingermarks on non-porous surfaces using anti-l-amino acid antibodies conjugated to gold nanoparticles. *Chemical Communications.* 2011;47(19):5602-4.

[68] Shenawi S, Jaber N, Almog J, Mandler D. A novel approach to fingerprint visualization on paper using nanotechnology: reversing the appearance by tailoring the gold nanoparticles' capping ligands. *Chemical Communications.* 2013;49(35):3688-90.

[69] Song K, Huang P, Yi C, Ning B, Hu S, Nie L, et al. Photoacoustic and colorimetric visualization of latent fingerprints. *ACS nano.* 2015;9(12):12344-8.

[70] Su C-H, Yu C-C, Cheng F-Y. Rapid visualization of latent fingermarks using gold seed-mediated enhancement. *Journal of nanobiotechnology.* 2016;14(1):75.

[71] Choi MJ, Smoother T, Martin AA, McDonagh AM, Maynard PJ, Lennard C, et al. Fluorescent TiO2 powders prepared using a new perylene diimide dye: Applications in latent fingermark detection. *Forensic Science International.* 2007;173(2-3):154-60.

[72] Theaker BJ, Hudson KE, Rowell FJ. Doped hydrophobic silica nano- and micro-particles as novel agents for developing latent fingerprints. *Forensic Science International.* 2008;174(1):26-34.

[73] Huang W, Li X, Wang H, Xu X, Liu H, Wang G. Synthesis of amphiphilic silica nanoparticles for latent fingerprint detection. *Analytical Letters.* 2015;48(9):1524-35.

[74] Zhang S, Liu R, Cui Q, Yang Y, Cao Q, Xu W, et al. Ultrabright fluorescent silica nanoparticles embedded with conjugated oligomers and their application in latent fingerprint detection. *ACS applied materials & interfaces.* 2017;9(50):44134-45.

[75] Zhang M, Ou Y, Du X, Li X, Huang H, Wen Y, et al. Systematic study of dye loaded small mesoporous silica nanoparticles for detecting latent fingerprints on various substrates. *Journal of Porous Materials.* 2017;24(1):13-20.

[76] Meng L, Ren Y, Zhou Z, Li C, Wang C, Fu S. Monodisperse silica nanoparticle suspension for developing latent blood fingermarks. *Forensic Sciences Research.* 2018:1-9.

[77] Chan WC, Maxwell DJ, Gao X, Bailey RE, Han M, Nie S. Luminescent quantum dots for multiplexed biological detection and imaging. *Current opinion in biotechnology.* 2002;13(1):40-6.

[78] Han M, Gao X, Su JZ, Nie S. Quantum-dot-tagged microbeads for multiplexed optical coding of biomolecules. *Nature biotechnology.* 2001;19(7):631.

[79] Galian RE, de la Guardia M. The use of quantum dots in organic chemistry. *TrAC Trends in Analytical Chemistry.* 2009;28(3):279-91.

[80] Bailey RE, Smith AM, Nie S. Quantum dots in biology and medicine. *Physica E: Low-dimensional Systems and Nanostructures.* 2004;25(1):1-12.

[81] Jamieson T, Bakhshi R, Petrova D, Pocock R, Imani M, Seifalian AM. Biological applications of quantum dots. *Biomaterials.* 2007;28(31):4717-32.

[82] Smith AM, Duan H, Mohs AM, Nie S. Bioconjugated quantum dots for in vivo molecular and cellular imaging. *Advanced drug delivery reviews.* 2008;60(11):1226-40.

[83] Hazarika P, Russell DA. Advances in fingerprint analysis. *Angewandte Chemie International Edition.* 2012;51(15):3524-31.

[84] Pan Z, Lai HL, Au FC, Duan X, Zhou W, Shi W, et al. Oriented silicon carbide nanowires: synthesis and field emission properties. *Advanced Materials.* 2000;12(16):1186-90.

[85] Wang YF, Yang RQ, Wang YJ, Shi ZX, Liu JJ. Application of CdSe nanoparticle suspension for developing latent fingermarks on the sticky side of adhesives. *Forensic Science International.* 2009;185(1-3):96-9.

[86] Algarra M, Jiménez-Jiménez J, Miranda M, Campos B, Moreno-Tost R, Rodriguez-Castellón E, et al. Solid luminescent CdSe-thiolated porous phosphate heterostructures. Application in fingermark detection in different surfaces. *Surface and Interface Analysis.* 2013;45(2):612-8.

[87] Wang YF, Yang RQ, Shi ZX, Liu JJ, Zhao K, Wang YJ. The effectiveness of CdSe nanoparticle suspension for developing latent fingermarks. *Journal of Saudi Chemical Society.* 2014;18(1):13-8.

[88] Moret S, Bécue A, Champod C. Cadmium-free quantum dots in aqueous solution: Potential for fingermark detection, synthesis and an application to the detection of fingermarks in blood on non-porous surfaces. *Forensic science international.* 2013;224(1-3):101-10.

[89] Cai K, Yang R, Wang Y, Yu X, Liu J. Super fast detection of latent fingerprints with water soluble CdTe quantum dots. *Forensic science international.* 2013;226(1-3):240-3.

[90] Cheng KH, Aijmo J, Ma L, Yao M, Zhang X, Como J, et al. Luminescence decay dynamics and trace biomaterials detection potential of surface-functionalized nanoparticles. *The Journal of Physical Chemistry C.* 2008;112(46):17931-9.

[91] Xu C, Zhou R, He W, Wu L, Wu P, Hou X. Fast imaging of eccrine latent fingerprints with nontoxic Mn-doped ZnS QDs. *Analytical chemistry.* 2014;86(7):3279-83.

In: Introduction of Forensic Nanotechnology … ISBN: 978-1-53616-040-6
Editors: R. K. Shukla and A. Pandya © 2019 Nova Science Publishers, Inc.

Chapter 4

NANOSENSORS FOR BODY FLUID IDENTIFICATION: RECENT ADVANCES AND FUTURE OPPORTUNITIES[1]

Anjani Prajapati[1] and Alok Pandya[1,†]
Department of Physical Sciences, Institute of Advanced Research,
Gandhinagar, Gujarat, India

ABSTRACT

Identification of body fluid at crime scenes is crucial to a fuller understanding of the circumstances pertaining to a crime. However, traditional body fluid identification methods are varied due to labor-intensive and technologically diverse techniques that do not permit the on-site identification of all body fluids such as blood, semen and saliva. Thus, there remains a need to develop on site conclusive identification of all forensically relevant biological fluids at the crime scene. Recent years, driven by the importance for forensic applications, nanosensors for the identification of human blood which has the potential to overcome the drawbacks of the current body fluid identification techniques. In this

---

It is hoped that this chapter will provide a glimpse into the multifaceted aspects of nanotechnology in body fluid identification.

† Corresponding Author's Email: alok.pandya@iar.ac.in.

chapter, we will discuss nanosensors and various use of nanosensors, body fluid identification, future opportunities and way ahead.

Keywords: body fluid, forensic, nanosensor

1. Understand Nanosensors

Nanotechnology is known to produce significant scientific and technological advances in the field of chemistry, biology, and physics for a wide range of applications [1]. It is already defined as the manipulation in science and engineering involving the design, synthesis, characterization, and application of materials and devices who is in nanometer scale. It is known that a nanometer is one billionth of a meter or three orders of magnitude smaller than a micron [2]. This nano dimension is playing a very important role in changing the world of science and technology. Herein especially, we will talk on the excellent application of Nano-intervention in science i.e., Nanosensors [3, 4]. Let us begin with our modern lives which rely on sensors to allow society to run smoothly. An example such as Sensors at shopping malls detect your presence and open doors to allow you to enter. Sensors measure the water level in your washing machine and ensure it doesn't overflow. Our central nervous system also work based on recognition of our activity. Similarly, Nanosensors work in much the same way, but in a more sensitive and specific way. It can also detect either minute particles or minuscule quantities of something. Therefore, Nanosensors is defined as nanomaterials based sensors which offer high sensitivity and specificity over sensors compared to traditional materials [5]. It is simply due to Nano-objects operate at a nanoscale as natural biological processes which allows functionalization with chemical and biological molecules, with recognition events that cause detectable physical and chemical changes. The difference between a sensor, biosensors and Nano biosensors is shown in Figure 1.

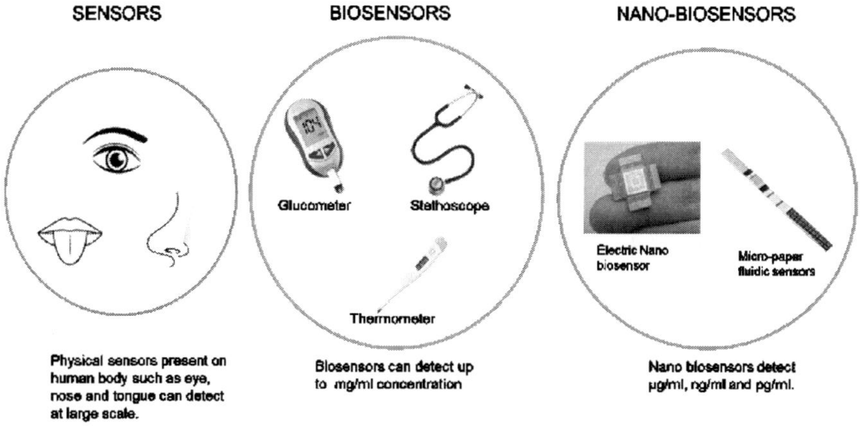

Figure 1. Difference between sensors, biosensors and Nano-biosensors.

These nanomaterials-based sensors have been designed to interact with cells and tissues at a molecular (i.e., subcellular) level with high functional specificity. Importantly, it allows a degree of coherence between technology and biological systems which were not attained previously [6]. Nanosensors are extensively developed by many institutions the forlast 10 years [7-10]. There have been several reports on advances in the research and development of Nanosensors for a number of different applications such as the medical field, national security, aerospace, integrated circuits, environment and many more [11-13]. Based on different applications for nanosensors, many different types of nanosensors, and a number of ways can develop them. There are a few challenges currently with the production of these nanosensors in terms of commercialization and sustainability which we will address later in this chapter. Basically, nanosensors can also be defined as "A chemical or physical or biosensor constructed using nanoscale components, usually microscopic or submicroscopic in size." Figure 2 is illustrating the three main types of nanosensors. Additionally, these sensors are very sensitive and can detect a single molecule at very low concentrations of a substance that could be potentially harmful. Now, we will understand the working of nanosensors.

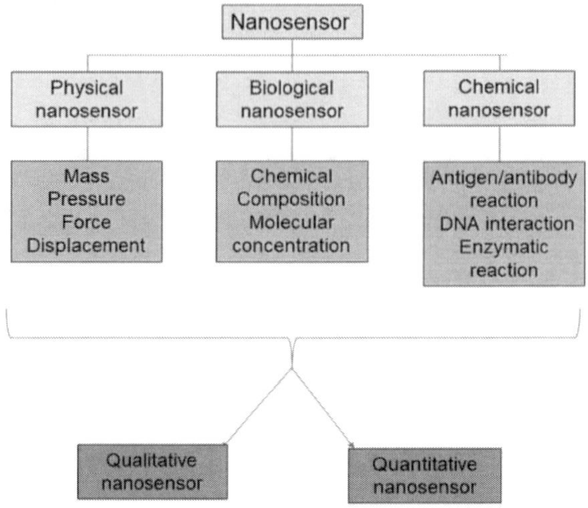

Figure 2. Types of nanosensors.

2. Working of Nanosensors

Now let us understand how nanosensors work for any type of application in sensing. A nanosensor consists of a sensitive layer that either contains recognition elements or covalently attached recognition elements to the transducer. The interaction between the target analyte and the receptor is designed to produce physicochemical changes that can be converted into a measurable effect [14]. Receptors are very important and provide specificity for sensor or biosensor technologies. It allows for binding of the specific analyte of interest to the sensor for the measurement with minimum interference from other components in sampling mixtures. Biological sensing elements can be either a biological molecular species (e.g., an antibody, an enzyme, a protein, or a nucleic acid) or a living biological system (e.g., cells, tissue, or whole organisms) which used as a biochemical mechanism for recognition [14]. Various readout techniques optical, electrochemical, or mass-sensitive can be used for detection in biosensors. Figure 3 is depicting the idea in general how a nanosensor works.

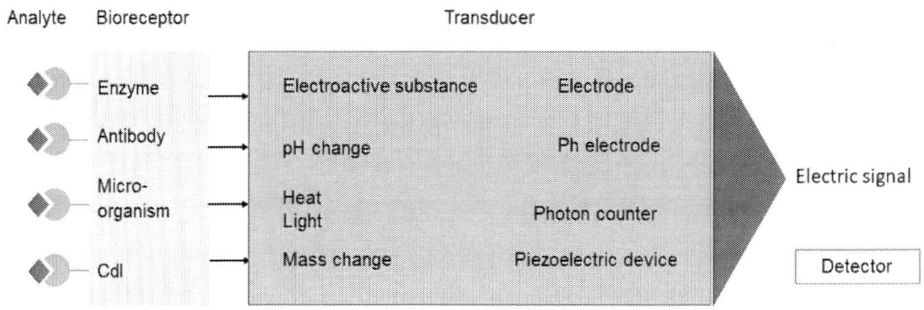

Figure 3. Schematic representation of the working of a nanosensor.

Majorly, two groups of receptor molecules form the Nanosensors, i.e., *Affinity-based Nanosensors* which are used to bind molecular species of interest, irreversibly and non- catalytically. Examples include antibodies, nucleic acids, and hormone receptors. And, *catalytic- based sensors* such as enzymes and microbiological cells recognize and bind a molecule of interest, followed by a catalyzed chemical conversion of that molecule to a product that is then detected [14]. Basically, it operates by charge transfer which occurs between molecules and sensitive material, resulting in an electrical and/or optical signal that is related to the type and number of the molecule.

3. Types of Nanosensors

Based on the analysis Nanosensors can be classified as follows:

3.1. Quantitative Nanosensors

The quantitative Nanosensor is summarized in three sections:

1) *Physical nanosensors*: Primarily, these sensors are used to measure magnitudes such as mass, pressure, force, or displacement. Their working principle is usually based on the fact that the electronic properties of both nanotubes and nanoribbons change when these

are bent or deformed [15]. For example, a CNT can be used to build a field-effect transistor (FET) nano in size, whose on/off threshold depends on the tube dimensions, shape, and temperature, amongst others. Local deformation of the tube creates a change in the on/off threshold voltage of the transistor. Starting from this simple principle, different types of nanoelectromechanical systems (NEMSs) have been proposed in the literature with different applications, such as pressure Nanosensors [15], force Nanosensors [15] or displacement Nanosensors [15].

2) *Chemical nanosensors*: The chemical nanosensors are utilized for measuring the concentration of an unknown gas, to know the chemical composition of a substance and for the detection of the target molecule. Commonly the chemical nanosensors works on the basic principle that the change in the electronic properties of CNTs and GNRs changes when different types of molecules are absorbed on top of them, which gradually increases or decreases the number of electron passes through the carbon lattice [15]. Same as in physical sensors, when a nanoribbon or nanotube is utilized in transistor composition, if there is a presence of target molecule, it will change the on/off threshold voltage of the transistor.
3) *Biological nanosensors:* These are used to monitor biomolecular processes such as antibody/antigen interactions, DNA interactions, enzymatic interactions or cellular communication processes, amongst others. A biological nanosensor is usually composed of (i) a biological recognition system or bioreceptor, such as an antibody, an enzyme, a protein or a DNA strain, and (ii) a transduction mechanism, e.g., an electrochemical detector, an optical transducer, or an amperometric, voltaic or magnetic detector [15]. There are mainly two subtypes of biological nanosensors based on their working principle: electrochemical biological nanosensors and photometric biological nanosensors. The electrochemical biological sensors work in a similar way to chemical nanosensors, but in this case, the change in the electronic properties of, for example, a CNT-based FET transistor, is induced either by: (i) A protein or any other chemical composite that binds itself to the functionalized nanotube. (ii) A specific antigen that

binds itself to an antibody glued to the nanotube. (iii) A single-stranded DNA chain that binds itself to another DNA chain which has been attached to the nanotube. Based on this principle, nanosensors able to detect lung cancer, asthma attacks, a different common viruses such as the influenza virus, or the parasite responsible for malaria, have been already successfully manufactured [16]. The second subtype of biological nanosensors is based on the use of noble metal nanoparticles and the excitation using optical waves of surface plasmons, i.e., coherent electron waves at the interfaces between these particles. Simply stated, the resonant frequency of the surface plasmons resulting from light irradiation changes when different materials are adsorbed on and in between the particles. This technique, known as localized surface plasmon resonance (LPSR), is the underlying principle behind many biological nanosensors. One of the main constraints of this sensing mechanism is the requirement of an external source of light and a device which is able to measure and compare different resonant frequencies of the particles. We believe that this can be overcome by means of coordination and communication among nanosensors. For example, nanosensors could locally irradiate the same particles with much lower power and measure the reradiated energy at different frequencies. The result of this operation could then be processed or forwarded to a data sink.

The application of biological nanosensors is in examining the biological process such as antigen/antibody immunoreactions, DNA interactivity, enzymatic reactions or any other cellular process. The configuration of biological nanosensors is mostly of (a) recognition of biological system or receptor such as an enzyme, DNA strand or antibody, and (b) a transduction mechanism which consists of an optical transducer, an amperometric, an electrochemical detector and voltaic and magnetic detector [16]. Based on their working principle, there are mainly two subcategories: 1) electrochemical biological nanosensors which work same as chemical nanosensors, but the difference is that the properties of CNTs and GNRs changes when specifically, a protein or an antigen or even DNA strains bind to the nanotube present in the sensor. These types of nanosensorshavebeen successful in the detection of different common viruses such as influenza, asthma attack, lung cancer and malaria [16]. 2) is the photometric biological nanosensor which is based on the activity of metal nanoparticles which changes colors when the target molecule has come in contact or has been absorbed on or between the particles. This property of the metal nanoparticle is called surface plasmon resonance which is the key technique by which photometric biological nanosensors detects the target biomolecule. In Figure 4 we illustrate the basic working principle of a) CNT-based FET transistor, b) physical nanosensor, c) chemical nanosensor and d) biological nanosensor.

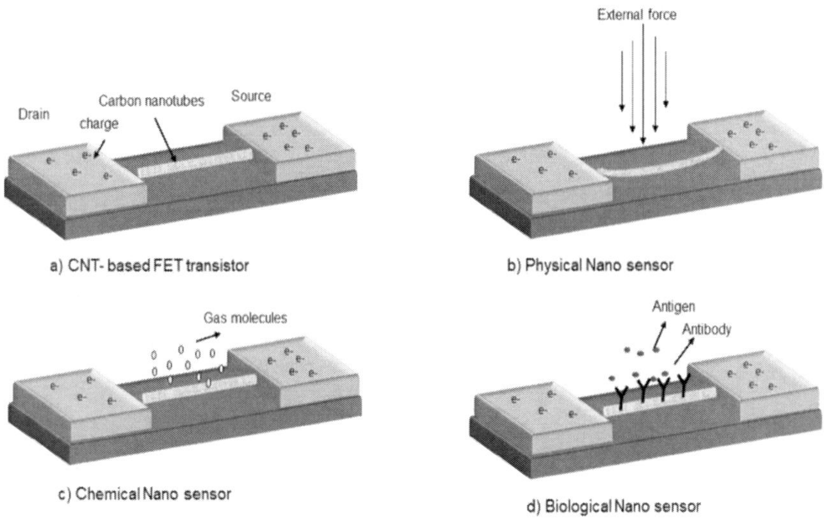

Figure 4. CNT-based FET transistor, b) Physical Nanosensor, c) Chemical Nanosensor and d) Biological Nanosensor.

3.2. Qualitative Nanosensors (Called Plasmonic Colorimetric Nanosensors)

Plasmonic refers to the phenomenon because of the free electrons oscillations on the metal surface under light stimulation [17]. When incident light has the same frequency as that of the surface electron oscillating against the attraction force to their positive nuclei, the resonance can be achieved, which is also described as surface plasmon resonance. Typically, two SPR modes have been explored: one is *propagating surface plasmon resonance* (PSPR) at the flat dielectric-metal interface; the other is *localized surface plasmon resonance* (LSPR) when those plasmons are highly confined to the surface of the colloidal nanoparticles or other nanostructures (e.g., two-dimensional nanostructured array). In plasmonic nanostructures, especially those made from gold and silver, the LSPR yields highly confined and enhanced electromagnetic fields on the boundary of nanoparticle surfaces, together with the appearance of a pronounced absorption peak in the visible and near-infrared frequency range. This implies that the nanoparticles size, shape, surface coating, and dielectric environments as well as the nanoparticles– chemical entities interaction can influence the oscillations, and these events can be transformed to the LSPR spectral change in either absorption/extinction or scattering and even color variation of the plasmonic nanoparticles solutions. These extraordinary plasmonic phenomena have given rise to a fast–developing field in colorimetric nanosensors, where the introducing of the targets induces the LSPR shift of plasmonic nanoparticles and is accompanied by the color change of the solution either by forming *cross-linking assembly/aggregate* or *transforming size/morphology of plasmonic nanoparticles [18].*

According to the different signal generation mechanism, these assays will be summarized into two categories of 1) *inter-nanoparticle distance–dependent colorimetric assay* and 2) *nanoparticles morphology/size–dependent colorimetric assay* (Figure 5).

1) *Inter-Nanoparticle Distance–Dependent Colorimetric Assay:*

   The aggregation of plasmon nanoparticles with appropriate sizes (diameter > 3.5 nm) will induce the surface plasmon coupling between the neighboring particles, along with the LSPR shift and visible color transition of nanoparticles solution [17]. For example, the AuNPs aggregation (or re-dispersion) events correspond to the tonality transition of the solution from red to blue (or from blue to red). Therefore, this inter-particle distance-dependent optical property of plasmonic nanomaterials provides a practical basis of the colorimetric assays for the sensing of any targets, such as metal ions, small molecules, macromolecules or living cells. Furthermore, we need to understand the driving force of such aggregations. So, molecular interactions driving the aggregates of plasmonic nanoparticles and their applications in molecular detection, which has been rarely discussed. Therefore, technically, the responsible molecular interactions such as various covalent/non-covalent forces, including *electrostatic interactions, hydrophobic forces, hydrogen bonding, and specific biological bonding* among plasmonic nanoparticles during aggregation are well explained and established.

2) *Nanoparticles Morphology/Size–Dependent Colorimetric Assays*:

Besides the inter-nanoparticle distance–dependent mechanism (i.e., target–triggered inter-particles cross-linking or aggregation) for colorimetric plasmon nanosensors, the precise tuning of the morphology/shape and size of plasmonic nanoparticles have also been applied for the colorimetric sensing in the target–mediated molecular events [17]. Since the growth of the nanoparticles can be tuned by many synthetic parameters, such as precursor, reducing agents, capping agents and others, there have been numerous reports on the controlling growth with some key parameters, i.e., very small change will result a totally different resulted nanoparticles, yielding differentiable features in the optical properties, particularly LSPR. Therefore, through tuning the kinetics of growth/etching of plasmonic nanoparticles, a variation of LSPR and solution color can be easily identified, and thereby the morphology/size–dependent properties can be controlled by some target–induced recognition events. In the light of the evolution approach in the size and morphology of nanoparticles, the related colorimetric nanosensors will be categorized into *target-induced plasmonic nanoparticles growth* and *plasmonic nanoparticles etching– dependent colorimetric assays. Target-*

induced plasmonic nanoparticles growth is the biomolecular–related reducing agents which have been chosen as the key parameter for controlling the growth of the nanoparticles. With this, hydrogen peroxide (H_2O_2) and other reducing agents can be quantified by the growth of the plasmonic nanoparticles–based sensing mechanism. For example, H_2O_2 can reduce $AuCl_4^-$ (or Ag^+)to Au (0) [or Ag (0], thus enlarging the seeding AuNPs (or AgNPs) and leading to an obvious LSPR change. *Nanoparticles etching– dependent colorimetric assays* are based on controllable chemical etching, which is one of the most vital strategies for controlling the nanostructure, has been used in electronics, catalysis and optic sensing applications, including in the plasmon-based colorimetric sensing systems. Usually, it involves the oxidation of plasmonic nanoparticles in the presence of the etchant, leading to the composition/morphology change of the nanoparticles, which can be naked-eye detected.

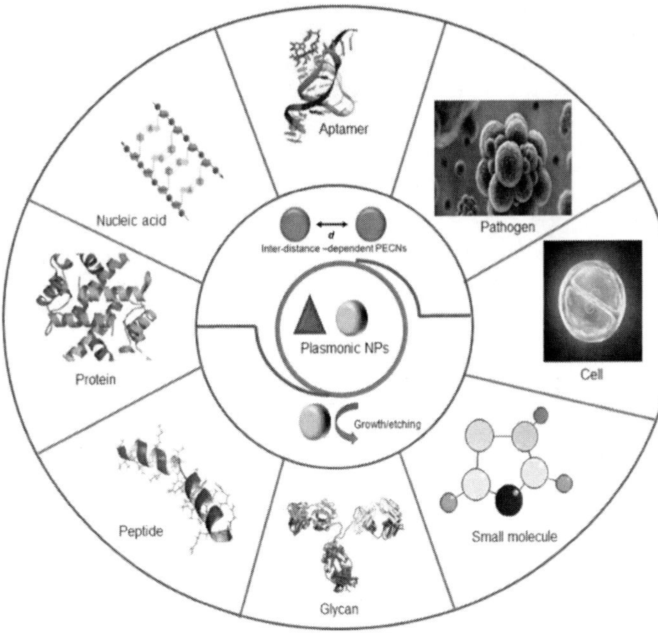

Figure 5. Schematic representation of inter-particle distance-dependent and size/morphology-dependent assays.

In summary, in this section, we understand about types and workings of nanosensors. We also understand what effective strategy should be to make ultrasensitive and selective nanosensor. Now we will see the about body fluids and about various reports of traditional technique and nanosensor to detect the body fluids.

4. Body Fluid as Forensic Specimen and Composition

During a forensic investigation, body fluid traces recovered at crime scenes are among the most important types of evidence [18]. These traces contain blood, semen, saliva, vaginal fluid, urine and sweat which can identify a suspect or victim as well as exonerate an innocent individual. Therefore, body fluid identification is an important component in the toolbox of forensic experts to carry out both at the crime scene and in the laboratory for many years. Most of the body fluid stains are present in small quantity, or sometimes are in the mixture and are invisible, so the investigation or the identification of these types of stains is not usually straightforward. Traditionally, the presumptive tests use enzymatic or chemical assays which are often limited in sensitivity or specificity, on the other hand, confirmatory tests are based on immunological or microscopic tests. Many of the early tests were incompatible with DNA profiling and consumed already limited biological material [18].

In forensic science, blood, saliva, menstrual blood, semen, and vaginal materials are of great interest in the investigation. Table 1 shows the different composition of different body fluids. For the identification of these body fluids, the test should be designed in such a way that it is easy to use at a crime scene or in the laboratory, it should be specific and sensitive and should also be nondestructive, should also allow the analysis by DNA profiling. The test should also be functional on different substrates as the body fluids tend to be deposited on different surfaces. DNA profiling has become sensitive with the progress of new multiplexes and there is a need for complementary body fluid identification tools.

Recent developments in the zone of body fluid identification have emerged in a number of new methodologies and approaches permitting 1) the conclusive detection of some body fluids for the first time, 2) development a nondestructive testing at the crime scenario for further analysis with fast and effective DNA profiling, and 3) the identification of a single cell by laboratory-based RNA and protein techniques [18]. Driven by the importance for forensic applications, body fluid identification methods have been extensively developed in the last 15 years. Firstly, we will understand the traditional approach to identify body fluids.

Table 1. Composition of body fluids:
Blood, semen, saliva, vaginal fluid, urine and sweat

Blood	Semen	Saliva	Vaginal fluid	Urine	Sweat
Hemoglobin	Prostate specific antigen	Amylase	Acid phosphatase	Urea	Urea
Erythrocytes	Spermatozoa	Lysozymes	Lactic acid	Creatinine	Lactic acid
Albumin	Choline	Mucin	Citric acid	Uric acid	Chloride
Glucose	Spermine	Thioc-yanate	Urea	Chlorine	Potassium
Immuno-globulins	Semenogelin	Glucose	Vaginal peptide	Tammhorsfall glycoprotein	Immuno-globulins
Fibrinogen	Zinc		Pyridine		

4.1. Traditional Spectroscopic Tests for Body Fluid Identification

There are various methods to identify the body fluids, but traditional techniques come along way for the primary or presumptive test as shown in Figure 6. A different group of forensic scientists has recorded their data for the body fluid identification test.

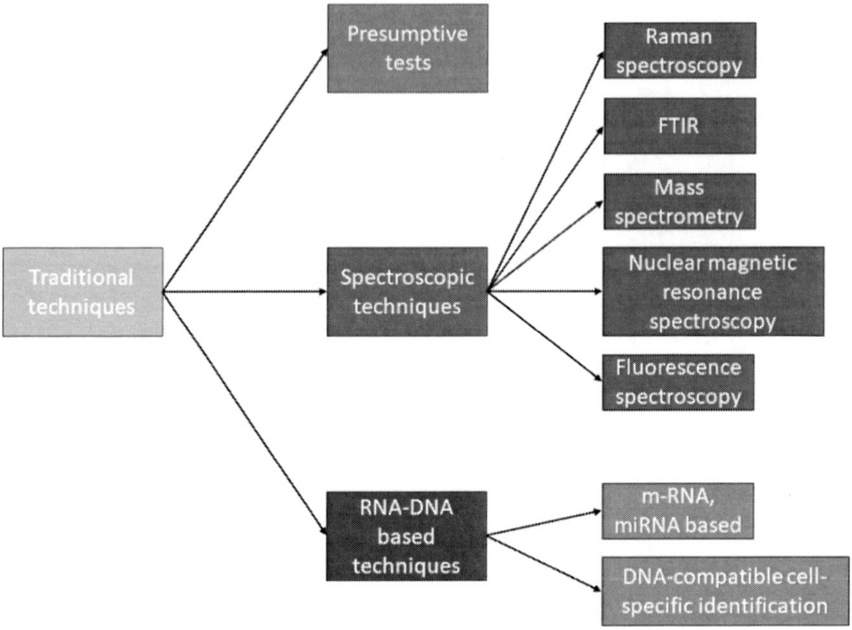

Figure 6. Traditional techniques used for the identification of body fluids.

Tyler Schlageter, et al. have reported their work for the identification of body fluid using Raman Spectroscopy. Human body fluids are of great interest in forensics, due to the possibility to extract their genetic information. At the moment, there is the need to develop a non-destructive, rapid and user-friendly method for the detection and identification of the body fluids usually found at crime scenes: blood, semen, vaginal fluid, saliva, sweat, and urine. The group has reported the spectroscopic techniques used or being researched on this topic, taking into account their advantages, limitations, and advances. Although UV-Vis light sources are used worldwide in forensic laboratories for the location of body fluids, they are not selective enough to be confirmatory. HSI, FTIR and Raman spectroscopy seems to be suitable for the identification and discrimination of body fluids, through comprehensive research about some unsolved aspects must be performed first [19].

Vitali Sikirzhytski, et al. have reported their work for the body fluid identification by the use of Raman Spectroscopy. The work was done by using confocal Raman Spectroscopy with a 785-nm excitation wavelength under controlled conditions. Detection and identification of blood, semen, and saliva stains are the most common body fluids encountered at a crime scene. This study targets the development of a nondestructive, confirmatory method for body fluid identification based on Raman spectroscopy coupled with advanced statistical analysis. Dry traces of blood, semen, and saliva obtained from multiple donors were probed using a confocal Raman microscope with a 785-nm excitation wavelength under controlled laboratory conditions. Results demonstrated the capability of Raman spectroscopy to identify an unknown substance to be semen, blood or saliva with high confidence [19].

The analysis of biological fluids at a crime scene or on an article of physical evidence can assist investigators in understanding the circumstances surrounding a violent crime as well as provide the link between a suspect, the victim and the crime scene through DNA analysis. Initial screening for biological fluid evidence typically involves scanning an area that is untreated or has been treated with fluid contrast enhancement reagents, with a specific wavelength of light while wearing colored goggles. Any contrast that is observed could indicate biological evidence. Once evidence is located, chemical tests are often performed to presumptively identify the unknown material. Both optical searching and chemical test techniques are reasonably reliable but present a unique set of challenges. These challenges include the need to work in a completely dark environment and the potential for sample adulteration/destruction due to the use of contrast enhancement reagents. These challenges, along with other difficulties associated with forensic casework (e.g., inhomogeneous samples located on non-ideal substrates, evidence recovered under non-ideal environmental conditions, the inability to appropriately obtain evidence at a crime scene, etc.) emphasize the necessity to modify existing techniques and instrumentation to better locate, capture, and document evidence.

In this study, ChemImage's HSI Examiner™ 1000 (Examiner), employing visible and near-infrared (NIR) reflectance hyperspectral imaging (HSI), was used to observe saliva, sweat, urine, blood and semen stains on a variety of natural and semi fabrics. The Examiner's capability for visualizing and presumptively locating biological stains on a variety of challenging substrates is demonstrated [20].

Blood is one of the most commonly encountered types of biological evidence found at scenes of violent crime and one of the most commonly observed fingerprint contaminants. Current visualization methods rely on presumptive tests or chemical enhancement methods. Although these can successfully visualize ridge detail, they are destructive, do not confirm the presence of blood and can have a negative impact on DNA sampling. A novel application of visible wavelength reflectance hyperspectral imaging (HSI) has been used for the detection and positive identification of blood-stained fingerprints in a non-contact and non-destructive manner on white ceramic tiles. The identification of blood was based on the unique visible absorption spectrum of hemoglobin between 400 and 680 nm. HSI has been used to successfully visualize ridge detail in bloodstained fingerprints to the ninth depletion. Ridge detail was still detectable with diluted blood to 20-fold dilutions. Latent blood stains were detectable to 15,000-fold dilutions. Ridge detail was detectable for fingerprints up to six months old. HSI was also able to conclusively distinguish bloodstained fingerprints [21].

Body fluid (BF) identification is a critical part of a criminal investigation because of its ability to suggest how the crime was committed and to provide reliable origins of DNA. In contrast to current methods using serological and biochemical techniques, vibrational spectroscopic approaches provide alternative advantages for forensic BF identification, such as non-destructive and versatility for various BF types and analytical interests. However, unexplored issues remain for its practical application to forensics; for example, a specific BF needs to be discriminated from all other suspicious materials as well as other BFs, and the method should be applicable even to aged BF samples. Herein, we describe an innovative modeling method for discriminating the ATR FT-IR spectra of various BFs, including peripheral blood, saliva, semen, urine, and sweat, to meet the practical demands described above. Spectra from unexpected non-BF samples were efficiently excluded as outliers by adopting the Q-statistics technique. The robustness of the models against aged BFs was significantly improved by using the discrimination scheme of a dichotomous classification tree with hierarchical clustering. The present study advances the use of vibrational spectroscopy and a chemometric strategy for forensic BF identification [22].

Blood, saliva, semen, and vaginal secretions are the main human body fluids encountered at crime scenes. Currently, presumptive tests are routinely utilized to indicate the presence of body fluids, although these are often subject to false positives and limited to particular body fluids. Over the last decade, more sensitive and specific body fluid identification methods have been explored, such as mRNA analysis and proteomics, although these are not yet appropriate for routine application. This research investigated the application of ATR FT-IR spectroscopy for the detection and discrimination of human blood, saliva, semen and vaginal secretions. The results demonstrated that ATR FT-IR spectroscopy can detect and distinguish between these body fluids based on the unique spectral pattern, combination of peaks and peak frequencies corresponding to the macromolecule groups common within biological material. Comparisons with known abundant proteins relevant to each body fluid were also analyzed to enable specific peaks to be attributed to the relevant protein components, which further reinforced the discrimination and identification of each body fluid. Overall, this preliminary research has demonstrated the potential for ATR FT-IR spectroscopy to be utilized in the routine confirmatory screening of biological evidence due to its quick and robust application within forensic science [23].

4.2. RNA and DNA Based Technologies

RNA analysis has become a reliable method of body fluid identification for forensic use. Previously, scientists have reported the development of four multiplex quantitative PCR (qRT-PCR) probes to discriminate four different body fluids (blood, semen, saliva, and vaginal secretion). While those makers successfully identified most body fluid samples, there were some cases of false positive and negative identification. To improve the accuracy of the identification further, Jong-Lyul Park, et al. have attempted to use multiple markers per body fluid and adopted the NanoString nCounter system instead of a multiplex qRT-PCR system. After measuring tens of RNA markers, they evaluated the accuracy of each marker for body fluid identification. The study showed that the identification of blood and semen was accurate enough whereas, for vaginal secretion and saliva no single marker was accurate. Thus, the group has designed a logistic regression model with multiple markers for saliva and vaginal secretion and achieved almost perfect identification. In conclusion, the NanoString nCounter has proven to be an efficient platform for identifying multiple RNA markers per body fluid and can be utilized for forensic RNA analysis [24].

S Ingold, et al. designed an in-house assay for the Ion Torrent PGM/S5 platform, which was the basis for a collective exercise within 17 EUROFORGEN and EDNAP laboratories, in order to test the efficiency of targeted mRNA sequencing to identify body fluids. The exercise was to examine the dried body fluid stains and, contributor of' own bona fide or samples of human origin, according to described protocols. The given primer sequence included 33 and 29 body fluid specified targets, respectively, to identify saliva, blood, vaginal secretion, semen, vaginal, menstrual blood, and skin. The results depicted medium to high count values in the body fluid or targeted tissue with mere to no counts in non-target body fluids. There was some inter-laboratory variability in reading counts, but overall the results of the laboratories were comparable in that highly expressed markers showed high read counts and less expressed markers showed lower counts. Hanson, et al. conducted a partial least squares (PLS) analysis on the facts, where blood, menstrual blood, semen, and saliva markers and samples clustered well. The results of this exercise support targeted mRNA sequencing as a well-founded body fluid identification technique that can be added to the repertoire of forensic MPS panels [25].

Identification of the biological samples present on the crime scene is important for getting knowledge about the surrounding circumstances.

However, classical protein-based techniques do not allow the identification of all body fluids and tissues, therefore, to overcome this challenge, molecular-based techniques or assays are in need to be developed for the identification of all forensically and thus molecular-based strategies for the conclusive identification of all forensically applicable biological fluids and tissues need to be developed. Messenger RNA (mRNA) profiling is an example of such a molecular-based proposition. Current mRNA-based body fluid identification methods involve capillary electrophoresis (CE) or quantitative RT-PCR (qRT-PCR) platforms, each with its own drawbacks. Both the assays need the usage of expensive fluorescently labeled probes. CE-based methods require amplification and detection steps individually which increases the analysis time. For qRT-PCR assays, only 2-3 markers can be utilized in a single reaction since each requires a different fluorescent dye. To simplify mRNA profiling assays, and reduce the time and cost of analysis, Hanson EK, et al. have developed single- and multiplex body fluid High-Resolution Melt (HRM) method for the identification of forensically applicable biological fluids and tissues. The integrated biomarkers include IL19 (vaginal secretions), IL1F7 (skin), ALAS2 (blood), MMP10 (menstrual blood), HTN3 (saliva) and TGM4 (semen). The HRM assays require only unlabeled PCR primers and a single intercalating fluorescent dye (Eva Green). Each specific body fluid marker can easily be detected by the presence of a different melt peak. Nonetheless, HRM assays have been extensively used for the detection of targeted isoforms for a unique gene target. However, the group has distinctively developed a multiplex HRM assays to allow the parallel detection of multiple targets every single reaction. The group also reported the development and performance estimation of the developed HRM assays. The results demonstrated the extensive use of HRM assays for rapid, and inexpensive, screening of biological evidence [26].

The recognition of human body fluids or tissues via mRNA-based profiling is very convenient for forensic investigations. Preceding studies have shown mRNA biomarkers are effective to detect the conception of biological samples. In this study, Yan Xu, et al. preferred 16 tissue biomarkers which are specific to assess their effectiveness and sensitivities for human body fluids and tissues recognition, including porphobilinogen deaminase (PBGD), hemoglobin beta (HBB) and Glycophorin A (GLY) for circulatory blood, protamine 2 (PRM2) and transglutaminase 4 (TGM4) for semen, 1 (HBD1) for vaginal secretion, matrix metalloproteinases 7 and 11 (MMP7 and MMP11) for menstrual blood, histatin 3 (HTN3) for saliva, statherin (STATH) for nasal secretion, dermcidin (DCD) for sweat and uromodulin (UMOD) for urine. The above mentioned 10 common forensic body fluids or tissues were utilized in the study. Based on the observation, a reverse transcription (RT) PCR multiplex assay, XCYR1, which includes 12 biomarkers (i.e., HBB, GLY, HTN3, PRM2, KRT4, MMP11, MUC4, DCD, UMOD, MMP7, TGM4, and STATH) and 2 housekeeping genes [i.e., glyceraldehyde-3-phosphate dehydrogenase (GAPDH) and 18SrRNA], was developed. This test was further evaluated with real case samples (with both single and mixture source) and it was confirmed that XCYR1 is potent to detect common body fluids or tissues (i.e., vaginal secretion, semen, saliva, vaginal secretion, menstrual blood, oral mucosa, nasal secretion, sweat, and urine) in forensic casework samples [27].

5. Nanobiosensors for Body Fluid Identification

As we understand that the identification of human blood at a crime scene can provide crucial information to an investigation whilst also providing a source of nuclear material which can be targeted for DNA profiling. The ability to characterize an unknown stain at the scene of the crime without having to wait for results from a laboratory is very critical in the development of forensic body fluid analysis. Driven by the importance for forensic applications, nanosensor for the identification of human blood which has the potential to overcome the drawbacks of the current body fluid identification techniques. Herein, we will discuss the nanosensor of each body fluid as follows:

5.1. Blood

Chemical tests for blood are sensitive but lack specificity. To overcome the issue, there have been developments of Nano-biosensors for the identification of the blood at the crime scene. First is the ABAcard Hematrace by Abacus Diagnostics, which has detected until the limit of 0.07μg hemoglobin/mL. Tests showed specificity to the human, higher primate, and ferret blood. Positive results were unaffected by a variety, of contaminants and under a range of conditions. The SERATEC HemDirect and RSID™ is also a nanobiosensor for the detection of human blood by immunochromatography technique in which first the sample is mixed with buffer provided with kit, the sample loaded onto the sample whole and the sample is allowed to run on the strips, after 5 minutes 2 band appears, one is the control band and the other is the test band. As shown in Figure 7, SERATEC Hem Direct kit depicts that on the appearance of both the bands, it is confirmed that the sample is human blood [28-31].

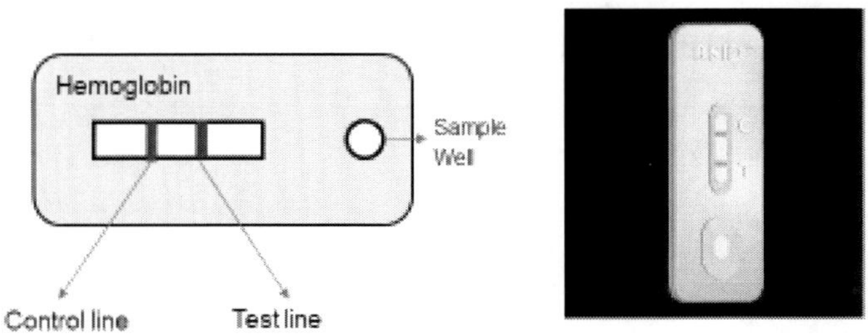

Figure 7. SERATECHemDirect and RSID™ detection kit.

5.2. Semen

For the detection of semen in a sample, the acid phosphatase reaction is a typical test in which the suspected seminal stain or an extract of it on a filter paper is reacted with a solution of the substrate, a monophenolic phosphoric acid or its ester, in acetate buffer of pH 5. The enzyme acid phosphatase hydrolyzes the substrate to the corresponding phenol and phosphate ion. The phenol formed is simultaneously coupled with a suitable diazonium salt as a chromogen to give characteristic colored dyestuff, which is a positive test for the presence of a seminal stain [32]. Seminogelin, which originates in seminal vesicles, has been used as a marker for the identification of semen in sample. A group of scientists, Satao l, et al. has identified the semen by using seminogelin with the help of one-step immunochromatographic assay. The group has reported their work for developing a system for the identification of semen. The system employed monoclonal antibody (mAb) and polyclonal antibody (pAb) against recombinant Sg-II (63 kDa), which has been synthesized in insect cells using baculovirus. The two antibodies bound with the seminal plasma motility inhibitor (SPMI; 14 kDa) as a final fragment peptide of Sg. The test stick is based on the sandwich technique using the above antibodies. When serial dilutions of seminal plasma were analyzed using this test stick, the intensity of a clear immunoreactive signal peaked at 2000-fold dilution. Thereafter, the signals decreased slowly but persisted up to 400,000-fold dilution. The Sg antigen was undetectable in saliva, urine, breast milk, serum or vaginal secretions. Also, the test stick shown did not react with animal semen samples, such as those from horses, dogs, swine, and bulls. When semen samples, diluted 100,000-fold from 100 men were tested, the Sg antigenic activity was detectable in all samples. In addition, the specificity and sensitivity of the test stick for identification of semen were demonstrated by comparative forensic studies [33]. The RSID-SEMEN is lateral flow immunochromatographic strip test, which uses two anti-human semenogelin monoclonal antibodies in a lateral flow format, to detect the presence of semenogelin. Semenogelin is a protein produced by the seminal vesicles and is responsible for the coagulum associated with ejaculate [34]. There are similar devices with the same principle as RSID, namely ABAcard p30 and SERATEC PSA [35].

5.3. Saliva

In forensic science examinations for the detection of saliva staining, a presumptive test called Phadebas is used. The test is designed to react with amylase, a digestive enzyme present in saliva [36]. For forensic identification of saliva, generally, there are assays for the enzymatic activity of alpha-amylase, an enzyme long associated with human saliva. Here, Old JB, et al. described the Rapid Stain IDentification (RSID-Saliva), a lateral flow immunochromatographic strip test that uses two anti salivary amylase monoclonal antibodies to identify the presence of salivary amylase. The group demonstrated that RSID-Saliva is accurate, reproducible, and highly sensitive for human saliva; RSID-Saliva detects less than 1 microL of saliva. The selectivity and the sensitivity of RSID-Saliva permit the examiner to sample a fragment of a suspected stain while keeping the rest of the majority for DNA-STR analysis. They also confirmed that RSID-Saliva detects saliva from a variant of materials (e.g., cigarette-butts, envelopes, bottles, and cans) and it does not react with blood, semen, urine, or vaginal fluid. RSID-Saliva is a convenient forensic test for evaluating which suspected items contain saliva and thus may yield a DNA profile.

5.4. Vaginal Fluid

Mostly vaginal discharge comprises of vaginal fluid, shedding vaginal and cervical cells, bacterias and cervical mucus. Most of the fluid in vaginal discharge is mucus produced by glands of the cervix. The mixture of vaginal fluid is secreted by cervical mucus, residual urine, and vaginal walls. In Spite of the fact, the vaginal secretions are not commonly found at most of the crime scenarios, though it is of great importance in the inquiry of sexual assault cases. As the composition of the vaginal fluids changes throughout the menstrual cycle, it is has been a challenge in the development of a technique or an assay for identifying the vaginal secretion on the crime scene. The most extensively used methods are based on the principle which uses periodic acid Schiff reagent for the identification of the glycogenated epithelial cells resulting in the cell being stained magenta is color. The intensity of the color is proportional to the concentration of the cells present. But the major drawback of this test is that the test becomes useless when the female of the category falls under postmenopausal or pre-pubescent because the vaginal secretion these females does not show glycogenated cells. Hannah holtkotter, et al. has reported the forensic validation of an immunochromatographic test that allows discrimination in forensic stains, the SERATEC PMB test, and its performance on real casework samples. As the PMB test is a binary test incorporates human hemoglobin and D-dimer detection and was developed for the detection of blood and menstrual fluid, both at the crime scenario and in the laboratory. This study revealed that the binary D-dimer/ hemoglobin assay detects the presence of human hemoglobin and recognizes samples having menstrual fluid by detecting the presence of D-dimers. Results also showed that the establishment of the new binary test is a considerable approach towards analyzing and interpreting evidence from sexual assault cases.

5.5. Urine

As urine mostly contains urea, proteins, hormones, urea and a wide range of metabolites, urine can be a potential body fluid for the detection in the investigation of cases like abuse or sexual assault. It is a challenge to establish a test for the identification of urine, as it is generally very dilute, although it typically fluoresces under an alternate light source. The presumptive test developed for the detection of urine is mostly based on the identification of urea present in the urine sample. The principle of this test is that is based on the enzyme urease which later breakdowns into urea and releases carbon dioxide and ammonia in the process. The most common is the method for detecting urine based on the presence of urea is the DMAC test which produces a pink or a magenta color if urea is present. Other than urea, creatinine is another content which is found to be in high concentration in the urine sample. For the detection of creatinine, the Jaffe test is utilized in which the picric acid is used which forms a red compound called creatinine picrate. The change in the color is proportional to the concentration of creatinine present. Similarly, the Salkowski test forms a blue color in the presence of urine by the reaction between creatinine and sodium nitroprusside when heated. Finally, the immunological RSID test has also been developed for the detection of urine, based on the presence of a urinary protein known as the Tamm-Horsfall protein (https://aboutforensics.co.uk/bodily-fluids-analysis/). There is no immunochromatographic test to date for the detection of urine.

5.6. Sweat

To date, there is no practical screening test for identifying sweat. Although DNA is frequently recovered and profiled from areas of clothing likely to contain sweat, little research has been undertaken. ELISA-based assays have been developed for the detection of sweat-specific protein G-81 and dermcidin but have not been widely adopted [19].

Conclusion and Way Ahead

Body fluid identification has been an important tool in criminal investigations for many years. The ability to detect and identify body fluids has also improved the last couple of years. A quick and easy-to-use presumptive test at the crime scene has been used to locate and indicate the body fluid of interest, with an available test in the laboratory and recently in the field. However, with the advent of on-site body fluid testing equipment such as the RapidHIT® system (IntegenX, Pleasanton, CA, USA), RSID™-saliva test, The RSID™-blood test for case samples, there is an increasing need for such sensitive and specific, nondestructive tests in the field. Portable spectrometers and nanodevices offer alternative solutions.

Laboratory-based methods for mRNA and miRNA profiling is developed using molecular biology techniques such as epigenetic approaches. Although challenges such as their sensitivity, selecting appropriately specific markers and relative ease of use are identified. It is a more complex problem when the sample is a mixture of body fluids with one or more at a low amount relative to the other which still needs to be addressed. The great promise is shown by recent research using graphene quantum dots, gold nanoparticle as a sensitive, specific approach to immunological testing, with the added advantage of being compatible with subsequent DNA on Chip process. The design and development of nanotools for body fluid identification is a huge opportunity for forensic researcher into the future. In all, the motivations behind such challenges would hopefully serve to increase public confidence in forensic science's ability to detect past crimes and prevent future ones.

ACKNOWLEDGMENTS

We gratefully acknowledge the funding from SERB, India under Young Scientist Start-up Grant (YSS/2015/000258).

REFERENCES

[1] Saha K, Agasti SS, Kim C, Li X, Rotello VM. Gold Nanoparticles in Chemical and Biological Sensing. *Chemical Reviews.* 2012;112 (5):2739-79.

[2] Howes PD, Rana S, Stevens MM. Plasmonic nanomaterials for biodiagnostics. *Chemical Society Reviews.* 2014;43 (11):3835-53.

[3] Pandya A, Bollapalli M. *Nanodevices for Early Diagnosis of Cardiovascular Disease: Advances, Challenges, and Way Ahead.* Nanobiotechnology: CRC Press; 2018. p. 69-103.

[4] Lim WQ, Gao Z. Plasmonic nanoparticles in biomedicine. *Nano Today*. 2016;11 (2):168-88.

[5] Akyildiz IF, Jornet JM. Electromagnetic wireless nanosensor networks. *Nano Communication Networks.* 2010;1 (1):3-19.

[6] Mendes PM. Cellular nanotechnology: making biological interfaces smarter. *Chemical Society Reviews.* 2013;42 (24):9207-18.

[7] Chowdury M, Walji N, Mahmud M, MacDonald B. based microfluidic device with a gold nanosensor to detect arsenic contamination of groundwater in Bangladesh. *Micromachines.* 2017;8 (3):71.

[8] Wilson DR, Routkevitch D, Rui Y, Mosenia A, Wahlin KJ, Quinones-Hinojosa A, et al. A Triple-Fluorophore-Labeled Nucleic Acid pH Nanosensor to Investigate Non-viral Gene Delivery. *Molecular Therapy.* 2017;25 (7):1697-709.

[9] Dave H, Bhandari N, Mod B, Shukla R, Pandya A. Multifunctional Silver-Cellulose Nanocomposite as a Promising Plasmonic Sensing Platform. *Journal of nanoscience and nanotechnology.* 2018;18 (8):5461-9.

[10] Ansari N, Lodha A, Pandya A, Menon S. Determination of cause of death using paper-based microfluidic device as a colorimetric probe. *Analytical Methods.* 2017;9 (38):5632-9.

[11] Sutariya PG, Pandya A, Lodha A, Menon SK. A simple and rapid creatinine sensing via DLS selectivity, using calix [4] arene thiol functionalized gold nanoparticles. *Talanta.* 2016;147:590-7.

[12] Lad AN, Pandya A, Agrawal Y. Overview of nano-enabled screening of drug-facilitated crime: a promising tool in forensic investigation. *TrAC Trends in Analytical Chemistry.* 2016;80:458-70.

[13] Neethirajan S, Ragavan V, Weng X, Chand R. Biosensors for sustainable food engineering: Challenges and perspectives. *Biosensors.* 2018;8 (1):23.

[14] Devreese JT. Importance of nanosensors: Feynman's vision and the birth of nanotechnology. *Mrs Bulletin.* 2007;32 (9):718-25.

[15] Lee J-H, Cho H-Y, Choi H, Lee J-Y, Choi J-W. Application of gold nanoparticle to plasmonic biosensors. *International journal of molecular sciences.* 2018;19 (7):2021.

[16] Tang L, Li J. Plasmon-based colorimetric nanosensors for ultrasensitive molecular diagnostics. *ACS sensors.* 2017;2 (7):857-75.

[17] Harbison S, Fleming R. Forensic body fluid identification: state of the art. *Res Rep Forensic Med Sci.* 2016;6 (2016):11-23.

[18] Schlagetter T, Glynn CL, Kammrath BW. *The use of Raman spectroscopy for the identification of forensically relevant body fluid stains.* 2017.

[19] Sikirzhytski V, Virkler K, Lednev IK. Discriminant analysis of Raman spectra for body fluid identification for forensic purposes. *Sensors.* 2010;10 (4):2869-84.

[20] Cadd S, Li B, Beveridge P, O'Hare WT, Campbell A, Islam M. The non-contact detection and identification of blood stained fingerprints using visible wavelength reflectance hyperspectral imaging: Part 1. *Science & Justice.* 2016;56 (3):181-90.

[21] Takamura A, Watanabe K, Akutsu T, Ozawa T. Soft and robust identification of body fluid using Fourier transform infrared spectroscopy and chemometric strategies for forensic analysis. *Scientific reports.* 2018;8 (1):8459.

[22] Orphanou C-M. The detection and discrimination of human body fluids using ATR FT-IR spectroscopy. *Forensic science international.* 2015;252:e10-e6.

[23] Park J-L, Park S-M, Kim J-H, Lee H-C, Lee S-H, Woo K-M, et al. Forensic body fluid identification by analysis of multiple RNA markers using NanoString technology. *Genomics & informatics.* 2013;11 (4):277.

[24] Ingold S, Dørum G, Hanson E, Berti A, Branicki W, Brito P, et al. Body fluid identification using a targeted mRNA massively parallel sequencing approach–results of a EUROFORGEN/EDNAP collaborative exercise. *Forensic Science International: Genetics.* 2018;34:105-15.

[25] Hanson EK, Ballantyne J. Rapid and inexpensive body fluid identification by RNA profiling-based multiplex High Resolution Melt (HRM) analysis. *F1000Research.* 2013;2.

[26] Xu Y, Xie J, Cao Y, Zhou H, Ping Y, Chen L, et al. Development of highly sensitive and specific mRNA multiplex system (XCYR1) for forensic human body fluids and tissues identification. *PloS one.* 2014;9 (7):e100123.

[27] Misencik A, Laux DL. Validation study of the seratec hemdirect hemoglobin assay for the forensic identification of human blood. *MAFS Newslett.* 2007;36 (2):18-26.

[28] Johnston S, Newman J, Frappier R. Validation study of the Abacus Diagnostics ABAcard® Hematrace® membrane test for the forensic identification of human blood. *Canadian Society of Forensic Science Journal.* 2003;36 (3):173-83.

[29] Schweers BA, Old J, Boonlayangoor P, Reich KA. Developmental validation of a novel lateral flow strip test for rapid identification of human blood (Rapid Stain Identification™-Blood). *Forensic Science International: Genetics.* 2008;2 (3):243-7.

[30] Frascione N, Pinto V, Daniel B. Development of a biosensor for human blood: new routes to body fluid identification. *Analytical and bioanalytical chemistry.* 2012;404 (1):23-8.

[31] Raju P, Iyengar N. Acid phosphatase reaction as a specific test for the identification of seminal stains. *J Crim L Criminology & Police Sci.* 1964;55:522.

[32] Sato I, Kojima K, Yamasaki T, Yoshida K, Yoshiike M, Takano S, et al. Rapid detection of semenogelin by one-step immunochromatographic assay for semen identification. *Journal of immunological methods.* 2004;287 (1-2):137-45.

[33] Chang ME. *A Comparison of Rapid Stain Identification Test for Semen (RSIDTM-Semen)*, Seratec® PSA Semiquant, and ABAcard® p30 Tests for the Forensic Identification of Seminal Fluid 2012.

[34] Pang B, Cheung B. Identification of human semenogelin in membrane strip test as an alternative method for the detection of semen. *Forensic science international.* 2007;169 (1):27-31.

[35] Mullen C. Amylase: Phadebas test, saliva. *Wiley Encyclopedia of Forensic Science.* 2009.

[36] Old JB, Schweers BA, Boonlayangoor PW, Reich KA. Developmental Validation of RSID™-Saliva: A Lateral Flow Immunochromatographic Strip Test for the Forensic Detection of Saliva. *Journal of forensic sciences.* 2009;54 (4):866-73.

In: Introduction of Forensic Nanotechnology ... ISBN: 978-1-53616-040-6
Editors: R. K. Shukla and A. Pandya © 2019 Nova Science Publishers, Inc.

Chapter 5

EXPLOSIVES DETECTION AND NANO TECHNIQUE

Anand Lodha[1,3,*], Niha Ansari[2] and Man Singh[3]
[1]Department of Forensic Science, Mody University of Science and Technology, Sikar, Rajasthan, India
[2]Institute of Forensic Science, Gujarat Forensic Science University, Gandhinagar, Gujarat, India
[3]School of Chemical Sciences, Central University of Gujarat, Gandhinagar, Gujarat, India

ABSTRACT

There is at present an urgent need for explosives detection in trace, because of recent terrorist activities. Almost all analytical chemical methods have been or are being applied for detection of explosives. Amongst them, nanotechnology-based systems were one of the successful methods for the trace level detection of explosive materials with high specificity in both solution phase and gas phase. This book chapter outlines different nanomaterials based techniques that might be utilized in detection of explosives, presents recent method developments, related legal issues and looks towards future developments.

* Corresponding Author's E-mail: dr.alodha@gmail.com.

Keywords: explosive, detection, nano materials, forensic, sensor, nanotechnology

1. INTRODUCTION

An explosion occurs as the result of a chemical reaction which releases a vast amount of physical or chemical energy accompanied by light, heat, and sound in a short space of time. The cause of an explosion could range from a simple accident in the home to a major terrorist incident. The construction of a bomb can be fairly simple, some requiring just a timer or remote control to create an initial spark, which ignites the bomb and triggers a blast.

Detonating cords, plastic tubes filled with powdered explosive, are common parts of most of the explosive devices. PETN (pentaerythritol tetranitrate) as powdered explosive in detonating cords, commonly used to link charges through the transmission of the detonation shock wave or as the explosive charge themselves. Commonly explosive devices contain boosters to amplify and transmit the shockwave to main charge from the detonator. Timers or remote control devices are commonly used as trigger to initiate the detonation. These start the initial charge, which then goes on to ignite the explosive and detonate the bomb.

Explosives can be classified into six broad classes based on their chemistry:

1) Aliphatic nitro compounds, such as nitromethane, hydrazine nitrate;
2) Nitroaromatic compounds, such as TNT, dinitrobenzene (DNB), hexanitrostilbene, picric acid;
3) Nitramines or nitrosamines, such as octogen (HMX) or RDX;
4) Nitrate esters, such as pentrite (PETN), ethylene glycol dinitrate (EDGN), nitroglycerine, and nitroguanidine (NQ);
5) Acid salts, such as ammonium nitrate; and

6) Organic peroxides, such as triacetone triperoxide (TATP) and hexamethylene triperoxide diamine (HMTD). Also known as home-made explosives (HMEs).

HMEs have very high vapor pressures because of the presence of volatile organic compounds (VOCs) such as acetone. Recently, the use of HME-based terrorism has grown rapidly as they can be manufactured easily. HMEs are extremely unstable and require special handling to avoid detonation. Most common explosives have extremely low vapor pressures at ambient temperature. The very low vapor pressures indicate that these molecules are extremely sticky and tend to adsorb to surfaces very easily. The sticky nature of explosive molecules on surfaces at room temperature also causes reduction of the molecules in delivery lines into sensor systems. Therefore, trace sampling of these families of explosives with very low vapor pressures is a challenging problem as they are found in trace amount.

A variety of improvised explosive devices (IEDs) are commonly encountered, their construction and mechanisms specific to their use.

Pipe Bombs – Consisting of a short length of metal of PVC pipe with both ends capped. Detonator or high or low explosives inserted through one side drilled end. Parts of this bomb survive the blast and possess evidences such as fingerprints and manufacturer details.

Vehicle Bombs – These have low explosive with a high explosive booster. The bombs are placed inside vehicles and shifted to target location.

Letter & Parcel Bombs – Contain a small amount of high explosive packed inside an ordinary letter or parcel. Forensic evidence shows presence as the devices rarely cause fires and so less evidence is destroyed.

Mines – These are covered devices containing high explosive, commonly used by the military. Mines are often hidden and detonated when touched. Much forensic evidence are obtained from the device.

Suicide Bombers – Suicide bombers, usually an individual wearing a vest of explosives or carrying a device, are primarily utilized by terrorist organizations. These types of bombs allow for the explosives to literally be walked into the target location and placed close enough so that maximum destruction can be caused. The identity of bombers can often be ascertained from CCTV footage and witness statements. Parts of the bomber may even be found, as the upper torso and head sometimes survive the blast and are found a distance away from the epicenter of the blast. Many terrorist organizations will publish or advertise the identity of the bomber after the event has occurred.

2. Sensor Development

Detection of trace quantity of explosive materials is of vital importance in homeland security, battlefield, and environmental monitoring and protection. Also, the most recent alarming increase in number and violence of terrorist bombings has made the task difficult for the detection of improvised explosive devices. To detect trace amounts of explosive compound is a challenging task because of a number of problems, such as variety of explosive materials available, cleverness of packaging, variability of venue, and low quantity of unexploded explosives, contaminated samples, and different sample collection procedures. The recent rise in global terrorism has required that the methods available should be able to detect explosives with high sensitive and at a low-cost.

In case of explosives in larger area, the use of specially trained sniffer dogs or detection dog is perhaps one of the firstborn methods of detecting explosives. These dogs are trained to use their excellent sense of smell to detect and locate the existence of even the smallest amount traces of explosive material. Other are electronic ‘sniffer’ devices which detect the volatile components emitted by explosives. The air is drawn through a filter, causing the components to collect where they can then be analyzed, giving some indication as to its composition. However some explosives, such as plastic explosives, are not easily detected by these devices, therefore canines may be more useful in this respect.

Various techniques may be utilized to detect explosive residues, either traces left behind by the explosive device or materials used in the production of the explosives. The major problem in detection of explosives is that explosive substances are not easily detectable using conventional approaches. Approaches to detect volatile substances, such as ion mobility spectrometry in conjunction with swabbing, whilst in routine use are largely only suitable for the screening of items of hand baggage. For the detection of traces of explosive substances in the air, issues related to the low vapor pressures of many explosives are only exacerbated further when these explosives are wrapped or packaged to avoid detection. The increasing use of peroxide-based explosives has led to much research into detection of this group of explosive substances, the issue being that many current chemical identification techniques are based on the nitrogen and carbon content of a substance for identification and this practice is not suitable for peroxide explosives. Selective and sensitive detection of explosives is very important in countering terrorist threats and securing homeland. Traditional security measures at airports includes the use of metal detectors to identify weapons concealed in conjunction by using X-ray machines for viewing the baggage contents. One major problem with explosive substances is that they are not easily detectable using conventional approaches and in this context, many terrorist groups have stop using metallic objects. For the detection of explosives, various spectroscopic techniques are been applied namely Ion mobility spectroscopy, Mass spectroscopy, Terahertz spectroscopy, Infra-red spectroscopy, Laser-induced breakdown, Raman spectroscopy and Cavity ring down spectroscopy, etc. Various other sensors such as chemical, electrochemical and luminescence sensors are also been utilized for the explosives detection. Most of these sophisticated instruments are bulky, expensive, and time-consuming procedures. Because of these limitations, such systems cannot be placed at strategic locations such as airports, government buildings, public places, transportation networks, infrastructures, and road networks with unpredictable vehicular and pedestrian traffic. Thus there is a need to develop miniature sensors that are sufficiently sensitive and selective, inexpensive, and acquiescent for bulk production.

Over recent years nanotechnology has become a rapidly expanding area of research by increasing techniques and knowledge. Nanomaterials possess unique characteristics at nanoscale level as compared to bulk materials, this property helps enhance sensitivity and selectivity. For example, carbon nanotubes display enhanced detection sensitivity as compared to sensors incorporating other carbon platform, as they have excellent electrical and thermal conductivity. Metal nanoparticle also provide enhanced sensitivities especially in electrochemical systems due to increased surface areas, catalytic effects and improving electrical conductivity. Metal nanoparticles are excellent optical sensors due to increased surface areas and a number of optical effects such as surface-enhanced Raman spectroscopy. Nano sensors have the potential to satisfy all the requirements for an effective platform for the trace detection of explosives.

3. TNT (2,4,6-Trinitrotoluene) Detection

The most commonly used high explosives in the last 100 years is 2,4,6-trinitrotoluene (TNT), which not only poses a security threat but also causes soil and water contamination. TNT is a suitable target analyte for chemical sensing devices. Analytical procedures for the trace detection of explosives typically involve the collection of vapor samples and their analysis by using a sensitive method. Although several sensitive and selective strategies are reported for TNT detection and other explosives, these methods are usually time-consuming, require bulky equipment, tedious sample preparation, and an expert operator. Furthermore, these available systems cannot be miniaturized and lacks the ability to perform an automated high-throughput analysis [1-5].

Current sensing methods of nitro-based explosives are gas and liquid chromatography [6], mass spectrometry [7], ion-mobility spectroscopy [8], enzymatic assays [9], and electrochemical detection [10], etc. Various authors have utilized nanotubes, nanowires, and nanoparticles for the detection of TNT showing very high selectivity and sensitivity.

3.1. Nanoparticles

Nanoparticles are nanomaterials having all three dimensions in nanoscale and are potential objects for the applications in a wide variety of fields of nanoscience including forensic science. Metallic nanoparticles as a class of material have many beneficial properties. Due to their small size more points of contact or surface area is available as compared to similar bulk compound. Metal nanoparticles have excellent catalysis properties and they can be applied when increased levels of binding are essential. Some metal nanoparticles possess unique optical qualities which make them well suited to act as sensors, while others have para-magnetism properties which are useful in medical imaging, especially in magnetic resonance imaging (MRI). Metal nanoparticles open up a new field in fundamental science and their potential technological applications. Metal nanoparticles (NPs) enhance the SPR response. The coupling between the localized surface plasmon (LSP) of the metal NPs and the surface plasmon polariton (SPP) associated with the thin metal surface leads to a deformation in the dispersion curve of the SPP, and to a shift in the SPR energy. These changes in the SPR curves resulting from the alteration of the dielectric properties of the metal surface helps in detection of various components present in the trace amount [11].

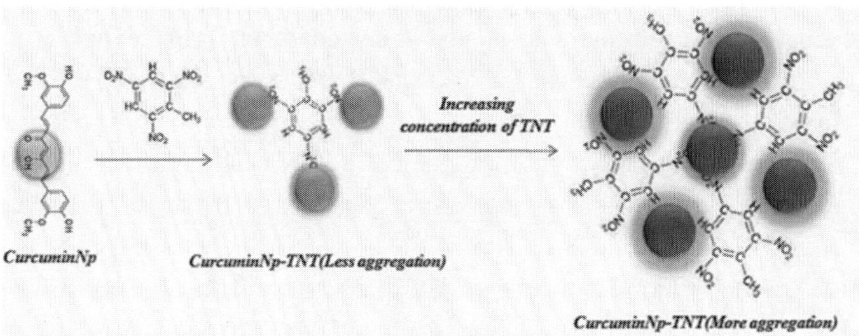

Figure 1. The schematic representation of the Pandya et al. work.

Pandya et al. designed an ultrasensitive nanocurcumin based nanomaterials surface energy transfer (NSET) probe for detection of TNT with excellent sensitivity of 1 nM which showed excellent selectivity over other nitro explosives via nanoaggregation. They found the largest fluorescent enhancement to date for sensing TNT up to 800 fold [12].

Noorhayati Idros et al. [13] developed colorimetric-based detection of TNT explosives using silica nanoparticles. They found that films of self-assembled functionalized nanoparticles are simple, selective, and label-free colorimetric sensing for a wide range of TNT and show reliable color changes after successful binding of TNT. Kawaguchi et al. [14] describe the use of gold nanoparticles on the immunosurface within a miniaturized SPR system. They found that the signal amplified four times for TNT and was absent in presence of only gold nanoparticle [14]. Wang et al. reported the use of Poly (guanine) - functionalized silica nanoparticles. The limit of detection was found to be ~100 ng L^{-1}. When these particles were complexed with anti-TNT coated magnetic beads using an analog of TNT, trinitrobenzene. When exposed to TNT, the analog was displaced and there was an increase of guanine in the solution [15]. Chen et al. described a chemo sensor based on pyrene functionalized Ru (ruthenium) nanoparticles, using 1-vinylpyrene (Ru = VPy) and 1-allylpyrene (Ru = APy). Both of these particles exhibited enhanced detection of nitro aromatic compounds in dimethylformamide as compared to monomeric pyrene. There was an increase in the response due to the nitration on the aromatic ring due to which TNT exhibited the greatest fluorescence quenching of the nitroaromatics under consideration [16]. Jiang et al. developed colorimetric sensor for TNT based on the donor-acceptor (D–A) interaction between TNT and primary amines (cysteamine). In the presence of TNT into the aqueous solution triggered the aggregation of these amine covered Au NPs and thus resulted in a color change from red to violet visually. Addition of as low as 114 pg L^{-1} TNT showed color change which can be visualized by the naked eye [17]. Dasary et al. described a SERS probe coated with gold nanoparticle. The system showed aggregation Meisenheimer complexes in water in the presence of TNT. The detection limit was fond to be 0.45 ng L^{-1} [18]. A similar technique was described by Yang et al. using functionalized silver nanoparticles. But their system showed lower detection limit that is 0.23 ng L^{-1} [19].

Xiaojuan Liu et al. designed a self-assembled gold octahedral array which uses a droplet evaporation process. This process produced a nanoparticle array with nanoscale inter-particle gaps which detects TNT by showing enhancement in the SERS through improved sensitivity and reproducibility [20]. Various authors have further worked in these area on factors like the effect of exposure time and temperature on the resulting signal [21, 22] as well as addressing false positives due to musk based perfumes of similar structure to TNT [23].

Michael Riskin et al. utilized π-donor cross-linked Au nanoparticles linked to the electrode for enhanced sensitive detection of TNT. They enhanced sensitive by tailoring πdonor-acceptor interactions between TNT and π-donor modified electrodes or π-donor cross-linked Au nanoparticles linked to the electrode. In one configuration a *p*-aminothiophenolate monolayer-modified electrode was utilized for the analysis of TNT with a sensitivity of 17 ppb (74 nM). In the second configuration, the cross-linking of Au NPs by oligothioaniline bridges to the electrode yields a functionalized electrode which detected TNT with a sensitivity of 460 ppt (2 nM). Further, they found that the imprinting of molecular TNT recognition sites into the π-donor oligoaniline-cross-linked Au nanoparticles yields a functionalized electrode with a sensitivity of 46 ppt (200 pM) [24]. This group also studied a bis-aniline cross-linked Au nanoparticles composite electropolymerized on Au surfaces for TNT detection via π donor-acceptor interactions. The detection limit for analyzing TNT by this method was found to be ~10 pM [25]. Daming Gao et al. reports a resonance energy transfer-amplifying fluorescence quenching at the surface of silica nanoparticles for the ultrasensitive detection of TNT in solution and vapor environments. The nanoparticle assembled arrays on silicon wafer were able to detect TNT up to ~1 nM [26].

Kawaguchia et al. [14] utilized immunosurface approach modified with Au nanoparticle as it has various advantages such as single step assay, label-free, low cost, minimized assay time, and less amount of antibody and Au nanoparticle. In order to obtain LSPR effect effectively, they utilized self-assembly technique based on thiolate–Au linkage to control the distribution of Au nanoparticle. In their work, they studied SPR signal amplification combined indirect competitive inhibition method with localized surface plasmon resonance (LSPR) effect. The TNT was detectable in the range of 10 ppt (10 pg/ml) to 100 ppb (100 ng/ml) from this immunosurface technique.

Pedrozo et al. detect TNT by using gold nanoparticles labeled SERS substrates. According to them, TNT detected up to a LDL of 1E−05 M simply by increasing the ionic strength of the gold nanoparticles colloidal suspension (through salt addition). Due to the increase in the ionic strength reduction of the anionic repulsive range (i.e., causes electric double layer compression) and displacement of anions from the gold nanoparticles surface occur thus offering room for TNT adsorption. On the other hand, under highly alkaline solutions (pH~12.5) of gold nanoparticles, TNT can be detected because of high alkaline conditions [27].

3.2. Nanotubes

Nanotubes possess unique properties, for example, their unique kinetic, optical and mechanical properties, their electrical properties as either conductors, insulators or semiconductors, depending on the structure and their high thermal conductivity in the axial direction whilst their thermal conductivity is low in the lateral direction [28].

Guo et al. developed an ionic liquid graphene nanosheet used with a glassy carbon electrode for TNT detection. The nanosheets were synthesized by using H. Yang et al. method [29] and used to detect the presence of TNT as it was getting electrochemically reduced. The detection limit of 4 μg L^{-1} in PBS was reported. When compared with plain glassy carbon electrodes or those functionalized with ionic liquid carbon nanotubes there was a distinct improvement in signal [30]. Wang et al. for the TNT detection developed glassy carbon electrodes by surface modification with multi-walled carbon nanotubes (MWCNT). The TNT get accumulated at the electrode surface due to the increased surface area due to the modification of the electrode. The LOD was found to be 0.6 μg

L^{-1} [31]. Further, it was discovered that this system performed better in a seawater matrix as compared to the synthetic NaCl solution. Chen et al. reported the use of semiconducting single-walled carbon nanotubes (SWCNT) as wearable chemical sensors. They firstly positioned the SWCNT in an aligned array on quartz using chemical vapor deposition and then transferring the array onto textile. Further, this coated textile section was then merged onto a chip carrier with an electrical and gas feed, thus converting it into a wearable transistor. In air, the LOD for TNT was found to be 8 ppb (80 ng L−1). The same group also reported the production of metal oxide nanowires with LOD 600 ng L^{-1} for TNT vapor [32]. Park et al. described a SWCNT based chemiresistive immunosensor for detection of nitroaromatic explosives trinitrophenyl (TNP). The SWCNT was first modified with trinitrophenyl (TNP), which was then functionalized with an anti-TNP single chain antibody. This antibody has a greater affinity for TNT and in the presence of TNT, a large change in the resistance of the system was noted. The detection range of 5 ng L^{-1} to 50 μg L^{-1} TNT in PBS and environmental water samples was reported [33].

Teresa Demeritte et al. developed thin film popcorn-shaped gold nanoparticle-decorated nanotubes acts as a sensitive hybrid SERS substrate. The metal nanoparticles were used as reactive sites for the binding with explosive materials. Due to the presence of a large surface area and very high aspect ratio, SWCNTs are selected as it proves to be a useful templates for the controlled attachment of popcorn-shaped gold nanoparticles which results in enhancement of the SERS signal. For the selective detection of TNT, the gold nanoparticle surface was modified with p-aminothiophenol. They were able to detect TNT material at a 100 femtomolar level [34]. Haibo Zhou et al. [35] reported in Figure 2 that TNT explosive lights up the Raman scattering of off-resonated p-aminobenzenethiol (PABT) through the rapid formation of a charge-transfer TNT-PABT complexing chromophore on the top-closed flexible silver nanotube array. The selectively detected for TNT was $1.5x10^{-17}$ M.

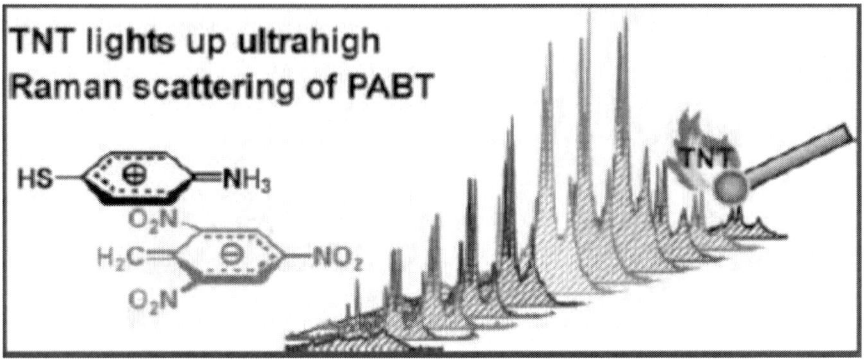

Figure 2. The schematic of et al. work.

Adem Yildirim et al. [36] synthesized mesotetrakis-p-carboxy-phenylporphyrin ($TCPPH_2$) doped silica nanotubes through peptide nanostructure templating for explosive sensing. The porous FSNT surface provides sensitive detection of TNT and DNT. The selectivity of the FSNT sensor was tested with other aromatic molecules namely nitrobenzene, toluene, xylene, benzoic acid, and di-hydroxybenzoic acid. Tae Hyun Kim et al. [37] developed selective nanocoatings by combining trinitrotoluene receptors bound to conjugated polydiacetylene polymers with single-walled carbon nanotube field-effect transistors (SWNTFET). The resulting sensors have detection limit up to 1 fM of TNT in real time, with excellent selectivity over various similar aromatic compounds.

3.3. Nanowires

For the detection of chemical and biological species, nanowire-based field-effect transistors (NW-FETs) prove to be have great potentiality as a sensors [38-41]. Recently, extensive work has been done on nanowire electrical devices for the detection of multiple biomolecular species of medical diagnostic relevance, such as DNA and proteins [42-44]. In case of nanowire sensors operated as field-effect transistors, the sensing mechanism involves the field-gating effect of charged molecules on the carrier conduction inside the NW [38-41]. Yoni Engel et al. [45]

demonstrate supersensitive, rapid, and label-free detection of TNT by the use of large-scale arrays of SiNW-FET devices. These devices were chemically modified with 3-aminopropyltriethoxy silane (monolayer of an amine-functionalized silane derivative). Through an acid–base-pairing interaction between TNT and amino ligands on the sensor surface the TNT molecules strongly binds to the surface of the nanowires. The SiNW devices empowers the detection of TNT with unprecedented sensitivities reaching sub-femtomolar ($<10^{-15}$ M) concentrations. Danling Wang et al. designed TiO2-(B) in the form of nanowires for the detection of 2,4,6-trinitrotoluene (TNT). The assembly of the TiO2-(B) nanowires by the hydrothermal method is a scalable and cost-effective technique having good practical approach. There finding proves that the reduction of cross-sensitivity to humidity through UV irradiation is an effective approach which can increase the performance of TiO_2-(B) nanowires sensor for the detection of explosive gas [46].

Sabahudin Hrapovic et al. combined metal nanoparticle with the nanotubes. They utilized metal nanoparticles together with multi-walled and single-walled carbon nanotubes solubilized in Nafion to form nano-composites for detection of TNT. The system showed reproducibility of the TNT signal within 7% (n = 8) from one electrode preparation to another, and the response signal was stable ((3.8% at 95% confidence interval) for 40 repeated analyses with 10 min of preconditioning. They analyzed TNT in tap water, river water, and contaminated soil [47].

3.4. Nanocrystals

Nanocrystals with high quantum yields have found a wider range of applications as a fluorescence sensors. Renyong Tu et al. synthesized Mn^{2+} doped ZnS nanocrystals with an amine-capping for the fluorescence detection of TNT by quenching the strong orange Mn^{2+} photoluminescence. The amine-capped nanocrystals can sensitively detect TNT down to 1 nM TNT in solution or atmosphere [48].

3.5. Molecular Imprinting

Chengguan Xie et al. reported the molecular imprinting at the walls of silica nanotubes for the recognition of TNT. They demonstrated that TNT templates were efficiently imprinted into the matrix of silica through the strong acid-base pairing interaction between TNT and 3-aminopropyltriethoxysilane (APTS). They synthesized TNT-imprinted silica nanotubes by the gelation reaction between APTS and tetraethylorthosilicate (TEOS), selectively occurring at the porous walls of APTS modified alumina membranes. Due to the ultrathin wall thickness of only 15 nm it provided greater capacity and faster kinetics of uptaking target species. The silica nanotube prove to be an ideal form of material for imprinting various organic or biological molecules toward applications in chemical/biological sensors and bioassay [49].

3.6. Nanospheres

Jian Yang et al. described the use of a blue emitting silafluorene–fluorene polymer, poly (silafluorene-(9,9-dimethyl-9Hfluorene) divinylene) (PSFF), coated on hollow silica nanoparticles as a chemosensor for the detection of aqueous solutions of TNT. The detection limit for TNT by silane–PSFF coated spheres is 200 ppb [50].

4. 2,4-Dinitrotoluene (DNT) Detection

4.1. Nanoparticles

Maung Kyaw Khaing Oo et al. detected direct DNT vapor at room temperature using gold nanoparticles (Au-NPs) synthesized with UV-assisted photo-chemical reduction method and subsequently self-assembled on the glass slide as the SERS substrates. They demonstrated real-time detection of DNT within 0.35 seconds by injecting vapor pulses through a glass capillary having inner surface was coated with Au-NPs [51].

4.2. Nanotubes

Adem Yildirim et al. utilized peptide nanostructure templating for synthesizing mesotetrakis-p-carboxy-phenylporphyrin (TCPPH2) doped silica nanotubes. The FSNT sensor detects DNT. They also analyzed $TCPPH_2$ doped nonporous silica thin film and silica nanoparticle sensors. The selectivity of the FSNT sensor was tested with nitrobenzene, toluene, xylene, benzoic acid and dihydroxybenzoic acid [36].

4.3. Nanocanals

Ko and Tsukruk demonstrated that DNT can be detected by fabricating a SERS substrate comprising of an array of nanocanals. Nanocanals can be formed by functionalizing the porous alumina membrane with the positively charged polymer poly (diallyldimethylammonium chloride, PDDA), having internal surface filled with gold nanoparticles [52].

4.4. Nanospheres

Jian Yang et al. described the use of a blue emitting silafluorene–fluorene polymer, poly (silafluorene-(9,9-dimethyl-9Hfluorene) divinylene) (PSFF) coated on hollow silica nanoparticles as a chemo-sensor for the detection of aqueous solutions of RDX. The detection limit of RDX by silane–PSFF coated spheres was found to be 800 ppb [50].

5. 2,4,6-TRINITROPHENOL (TNP)

5.1. Nanosphere

Yingxin Ma et al. utilized 8-hydroxyquinoline aluminum (Alq3) to form a fluorescent composite nanosphere for the TNP detection. Nanocomposites showed fluorescence quenching by addition of TNP and thus indicating that fluorescence intensity is negatively proportional to the TNP concentration. The detection limit was found to be 50 ng/mL for TNP, independent of interference of other nitroaromatic explosives, including TNT, DNT, and NB. Further, they converted the system into paper sensor-based detection of TNP by immersing the fluorescent filter paper into the nanosphere colloidal solution for the TNP detection [53].

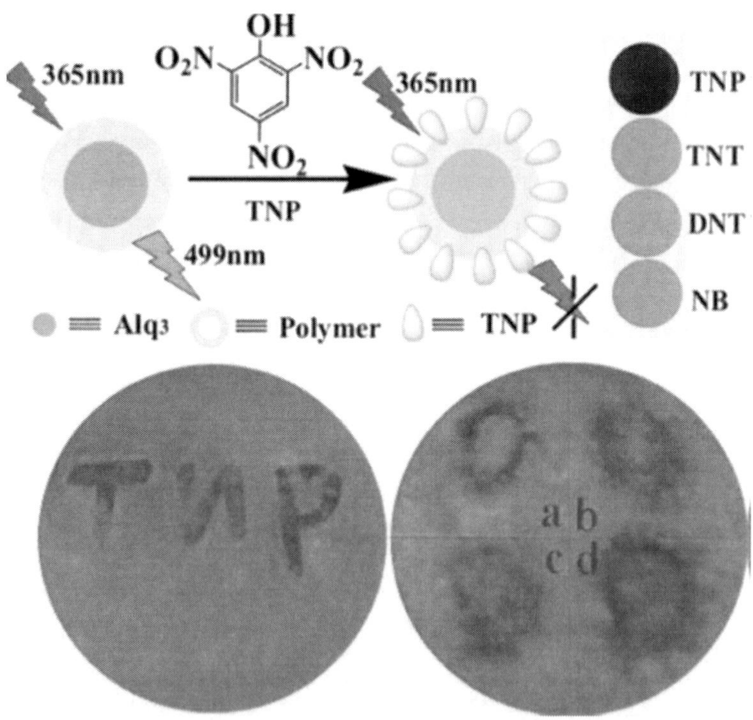

Figure 3. The schematic representation of Yingxin Ma et al. work.

6. Tri-Acetone Tri-Peroxide (TATP)

6.1. Nanotubes

Still, there is a critical need to develop an efficient, reliable and highly selective sensor for the detection of improvised non-military explosives. Subarna Banerjee et al. describes the utilization of functionalized titania nanotube arrays for the detection of improvised organic peroxide explosives. Tri-acetone tri-peroxide (TATP) forms complexes with titania nanotube arrays as a results of which the electron state of the nanosensing device get affected, which is later signaled as a change in current of the overall nanotube material. The response observed was rapid and a signal of five to eight orders of magnitude was noticed. These sensors can be converted into hand-held miniaturized devices [54].

7. Aromatic Nitro Compounds

7.1. Nanocrystals

Germanenko and co-workers measure the quenching of Si nanocrystals (λ_{EX}=355 nm, λ_{EM}=590–720 nm) in a methanol suspension. They studied quenching rate constants from Stern-Volmer analyses for 3,5-dinitrobenzonitrile, 4-nitrophthalonitrile, 1,4-dinitrobenzene, 4-nitrobenzonitrile, 2,3 dinitrotoluene, 3,4-dinitrotoluene, 2,4-dinitrotoluene, and 2,6-dinitrotoluene and found it in the range of 106-107 M^{-1} s^{-1}. Quenching of Si nanocrystals is very promising for explosive detection [55].

7.2. Nanotubes

Kose and coworkers studied carbon nanotubes functionalize with a poly (propionyl ethylenimine-ethylenimine) copolymer and report fluorescence (λ_{EX}=400 nm, λ_{EM}=510 nm). They found that the nanotube luminescence emissions are sensitive to the presence of nitroaromatic compounds. These effective luminescence quenching presents new opportunities for the use of carbon nanotubes in the continuing development of novel optical sensors [56].

7.3. Nanofibrous Film

Yufei Yang et al. developed electrospun nanofibrous film sensor based on tetrakis(4 methoxyl-phenyl)porphyrin (TMOPP) and polystyrene which showed favorable fluorescence sensing for four nitro aromatic explosives. The produced sensor showed high sensitivity as well as satisfactory reproducibility, reversibility, and selectivity. Porogen treatment produced secondary porous structures within the electrospun nanofibers and thus enhanced the sensitivity toward the nitro aromatic explosives. This work provides a facile and promising strategy to construct portable optical devices for the detection of nitro aromatic explosive [57].

7.4. Nanoribbons

Yanke Che et al. developed ultrathin nanoribbons from an amphiphilic electron donor-acceptor (D-A) supramolecule comprising perylene tetracarboxylic diimide as the backbone scaffold to enforce the one-dimensional intermolecular assembly via strong π-stacking. The photo generated electrons within the nanoribbon efficiently trapped by the adsorbed oxygen molecules or other oxidizing species, lead to depletion of the charge carriers and thus the electrical conductivity of the nanoribbon. Combination of this sensitive modulation of conductivity with the unique features intrinsic to the nanoribbon morphology enables the detection of nitro-based explosives [58].

8. 2,4,6-TRINITROPHENYLMETHYLNITRAMINE (TETRYL)

8.1. Nanoparticles

Yolanda Salinas et al. synthesized bi-functionalized silica nanoparticles via the independent anchoring of binding sites (amines) and reactive units (thiols). They utilized nanoparticles as a simple probe for the chromo-fluorogenic detection of Tetryl. In the absence of Tetryl, the thiol located on the surface of the nanoparticles reacts with a squaraine dye, inducing the bleaching of the solution and in the presence of Tetryl, they lead to the formation of charge transfer complexes between the explosive and amines grafted on the surface thus inhibiting the thiol–squaraine reaction [59].

9. RESEARCH DEPARTMENT EXPLOSIVE (RDX)

9.1. Nanoparticles

Larger sized gold nanoparticles are considered best as they have tendency to form aggregates. Various authors have demonstrated that aggregates of nanoparticles offer significant SERS enhancements when compared to individual nanoparticles [60]. Hatab et al. reported the application of cetyl-tri-methyl-ammonium bromide (CTAB) stabilized gold nanoparticles labeled on SERS substrates for the detection of RDX as low as 0.15 mg L^{-1}. The authors were able to obtain a SERS enhancement factor of 6E + 04 for RDX [61].

An imprinted composite coating of bisaniline–gold nanoparticles was developed using Kemp's acid as an imprinting molecule upon a gold surface for the detection of RDX. This technique used surface plasmon resonance (SPR) which reveals changes in the refractive index occurring due to adsorption or desorption at the surface. Detection of RDX was possible at 2.6 pg L^{-1} in HEPES buffer due to the formation of donor-acceptor complexes between RDX and the bisaniline bridging units [62].

As previously mentioned, detection of peroxide-based explosives can be complex due to these compounds containing neither the nitro groups nor aromatic behaviors traditionally used for analysis [63]. Banerjee et al. suggested the use of a one-dimensional titania (TiO_2) nanotube array where metal ions, in this case, zinc (Zn^{2+}), are coated upon the surface. When present in argon the TATP vapor bonded to the metal ions and the signal current output, as detected by a potentiostat, increased by between 3 and 5 orders of magnitude although this was found to be prone to saturation for TATP in air under ambient conditions [64]. The use of both carbon nanotubes and metallic nanoparticles were described by Hrapovic et al. for the electrochemical detection of nitroaromatic compounds. The nanocomposite was composed of a SWCNT with copper nanoparticles within a Nafion matrix, selected for their stability, reproducibility, low background current and prominent peak display with cyclic voltammetry and anodic stripping voltammetry, ASV. A detection limit of 1 $\mu g\ L^{-1}$ was reported in tap water, river water and washed contaminated soil samples [65].

Chang et al. reported a SERS-active nanoporous substrate based on porous alumina membranes combined with mixed nanoclusters comprising gold nanorods and gold nanoparticles. The plasmon resonance from hot spots produced by the above system photodecompose molecules of HMTD enabling analysis of the fragments. The LOD has been interpolated as 2 pg HMTD in 10 μL of acetonitrile [66].

Conclusion and Future Aspects

This review can hint at the exposure of this new area of research. The potential integration of all laboratory operations onto a single platform is perchance the most interesting development. It may further minimize laboratory contamination issues. Advanced capabilities such as enhanced SNP typing, quantum dot detection, and single-molecule sequencing may also transform this field by providing new commands for data collection and enhanced sensitivity for detection of trace DNA. Eventually, detection and sequencing of single DNA molecules left at a crime scene may not be far away.

Traditional DNA analysis is very complex, involving several steps and different instruments. However, remarkable advances have been possible in microfluidic systems in recent years in the field of forensic science and DNA analysis. Microfluidic systems, employing small volumes of fluids moving around the miniature cartridges, have the potential to deliver DNA readouts within few minutes. Developers of the microfluidic systems have introduced a paradigm shift for forensic DNA analysis by making these systems more accessible for a number of applications such as detection of drug molecule, explosive detection, and forensic relevant biomolecule identification. Although there are many more details to work out to improve the efficiency of microfluidic systems, researchers hope that the time invested in validating microfluidic measurements will pay off in the long rout. The future holds many promises for law enforcement and the communities they serve. There will be nanotechnology tools available for the detection of crime, investigating crimes, surveillance, and tracking tools for those citizens with disabilities that require the assistance of others to track their whereabouts. Preparing for our future today is the responsibility of our law enforcement leaders. They must know what technology will be available to protect their communities and know how they will acquire it. Our future promises a safer community through the use of nanotechnology. Our duty is to be trained, have set standards and protocols for the ethical usage of this emerging technology, and continually improving our tactics to solve crime and protect the public. Is law enforcements use of nanotechnology science fact or science fiction? It is, in fact, science reality and law enforcement must be prepared. While writing this critical review, we have noticed that only few people have been preferentially worked on nanotechnology-based research. Since not much work has been carried out on this area. We hope that after reading this critical review, researchers around the world will be motivated to enter into unknown territories of forensic evidence examination through nanotechnology. There is plethora of areas whose smart nanotechnology is yet to be studied. We also note the chosen use of certain nanomaterials, mainly carbon nanomaterial-based electrochemical or lab on chip based sensing of drugs and other small reliant molecule using nanotubes as well as gold nanoparticles (NPs) to prepare chemically modified electrodes, whilst nanoparticles made from silver, platinum, palladium, or metal oxides have not been employed vastly. Employing such nanoparticles as modifiers will predictably form a highly interesting topic for future research in the field of nanomaterial-based excellent and instant investigation to solve crimes.

REFERENCES

[1] Meaghan E. Germain and Michael J. Knapp. (2009) "Optical explosives detection: from color changes to fluorescence turn-on." *Chem Soc Rev*, 38(9):2543-55.

[2] David S. Moore (2004) "Instrumentation for trace detection of high explosives." *Rev. Sci. Instrum.,* 75:2499-512.

[3] Larry Senesac, Thomas G. Thundat (2008) "Nanosensors for trace explosive detection" *Mater. Today,* 11: 28-36.

[4] Jehuda Yinon (2003) "Peer Reviewed: Detection of Explosives by Electronic Noses" *Anal. Chem.,* 75(5), 99A-105A.

[5] Suman Singh (2007) "Sensors—An effective approach for the detection of explosives" *J. Hazard. Mater.* 144:15-28.

[6] Marianne E. Walsh (2001) "Determination of nitroaromatic, nitramine, and nitrate ester explosives in soil by gas chromatography and an electron capture detector" *Talanta.* 54(3):427-38.

[7] Christopher Mullen, Michael J. Coggiola, Harald Oser (2009) "Femtosecond laser photoionization time-of-flight mass spectrometry of nitro-aromatic explosives and explosives related compounds" *J American Soc Mass Spectr,* 20(3), 419–29.

[8] R. G. Ewing, R. G. Atkinson, G. A. Eiceman, G. A. Ewing. (2001) "A critical review of ion mobility spectrometry for the detection of explosives and explosive related compounds" *Talanta.*, 54(3):515-29.

[9] Sarlauskas J, Nemeikaite-Ceniene A, Anusevicius Z, Miseviciene L, Maroziene A, Markevicius A, Cenas N. (2004) "Enzymatic redox properties of novel nitrotriazole explosives implications for their toxicity" *Z Naturforsch C.*, 59(5-6):399-404.

[10] Joseph Wang, *Electronic Noses & Sensors for the Detection of Explosives*, 159 of the series NATO Science Series II: Mathematics, Physics and Chemistry, 131-42.

[11] Michael Riskin, Ran Tel-Vered, Oleg Lioubashevski, and Itamar Willner (2009) "Ultrasensitive Surface Plasmon Resonance Detection of Trinitrotoluene by a Bis-aniline-Cross-Linked Au Nanoparticles Composite," *J. Am. Chem. Soc.*, 131(21), 7368–78.

[12] Alok Pandya, Heena Goswami, Anand Lodha and Shobhana K. Menon (2012) "A novel nanoaggregation detection technique of TNT using selective and ultrasensitive nanocurcumin as a probe," *Analyst*, 137: 1771-4.

[13] Noorhayati Idros, Man Yi Ho, Mike Pivnenko, Malik M. Qasim, Hua Xu, Zhongze Gu, Daping Chu (2015) "Colorimetric-Based Detection of TNT Explosives Using Functionalized Silica Nanoparticles," *Sensors*, 15, 12891-905.

[14] T. Kawaguchi, D. R. Shankaran, S. J. Kim, K. Matsumoto, K. Toko, N. Miura, (2008) "Surface plasmon resonance immunosensor using Au nanoparticle for detection of TNT" *Sens. Actuators B: Chem.* 133(2): 467–72.

[15] Jun Wang, Guodong Liu, Hong Wu, Yuehe Lin (2008) "Sensitive electrochemical immunoassay for 2,4,6-trinitrotoluene based on functionalized silica nanoparticle labels" *Analytica Chim. Acta* 610:112–8.

[16] Wei Chen, Nathaniel B. Zuckerman, Joseph P. Konopelski and Shaowei Chen (2010) "Pyrene-Functionalized Ruthenium Nanoparticles as Effective Chemosensors for Nitroaromatic Derivatives" *Anal. Chem.* 82: 461–5.

[17] Y. Jiang, H. Zhao, N. Zhu, Y. Lin, P. Yu, L. Mao, (2008) "A simple assay for direct colorimetric visualization of trinitrotoluene at

picomolar levels using gold nanoparticles" *Angew. Chem. Int. Ed. Engl.* 47(45): 8601–4.

[18] Samuel S. R. Dasary, Anant Kumar Singh, Dulal Senapati, Hongtao Yu and Paresh Chandra Ray (2009) "Gold Nanoparticle Based Label-Free SERS Probe for Ultrasensitive and Selective Detection of Trinitrotoluene" *J. Am. Chem. Soc.* 131: 13806–12.

[19] Liangbao Yang Liang Ma Guangyu Chen Jinhuai Liu Prof. Zhong-Qun Tian (2010) "Ultrasensitive SERS Detection of TNT by Imprinting Molecular Recognition Using a New Type of Stable Substrate" *Chem. Eur. J.* 16: 12683–93.

[20] Xiaojuan Liua, Liang Zhao, Hao Shen, Hongxing Xu, Lehui Lu, (2011)"Ordered gold nanoparticle arrays as surface-enhanced Raman spectroscopy substrates for label-free detection of nitro explosives," *Talanta* 83:1023–9.

[21] X. Fang, S. R. Ahmad, (2009) "Detection of explosive vapour using surface-enhanced Raman spectroscopy" *Appl. Phys. B,* 97:723–6.

[22] Hainer Wackerbarth, Christian Salb, Lars Gundrum, Matthias Niederkrüger, Konstantin Christou, Volker Beushausen, and Wolfgang Viöl, (2010) "Detection of explosives based on surface-enhanced Raman spectroscopy" *Appl. Opt.* 49 4362–6.

[23] Hainer Wackerbarth, Christian Salb, Lars Gundrum, Matthias Niederkrüger, Konstantin Christou, Volker Beushausen, and Wolfgang Viöl (2010) "Challenge of false alarms in nitroaromatic explosive detection—a detection device based on surface-enhanced Raman spectroscopy" *Appl. Opt.* 49:4367–71.

[24] Michael Riskin, Ran Tel-Vered, Tatyana Bourenko, Eran Granot, and Itamar Willner, (2008) "Imprinting of Molecular Recognition Sites through Electropolymerization of Functionalized Au Nanoparticles: Development of an Electrochemical TNT Sensor Based on π-Donor-Acceptor Interactions," *J. Am. Chem. Soc.* 130:9726–33.

[25] Michael Riskin, Ran Tel-Vered, Oleg Lioubashevski, and Itamar Willner, (2009) Ultrasensitive Surface Plasmon Resonance Detection of Trinitrotoluene by a Bis-aniline-Cross-Linked Au Nanoparticles Composite, *J. Am. Chem. Soc.*, 131:7368–78.

[26] Daming Gao, Zhenyang Wang, Bianhua Liu, Lin Ni, Minghong Wu, Zhongping Zhang, (2008) "Resonance Energy Transfer-Amplifying Fluorescence Quenching at the Surface of Silica Nanoparticles toward Ultrasensitive Detection of TNT," *Anal. Chem.* 80:8545–53.

[27] Primera-Pedrozo, O. M., Soto-Feliciano, Y. M., Pacheco-Londoño, L. C. and Hernández-Rivera, S. P., (2009) "Detection of High Explosives Using Reflection Absorption Infrared Spectroscopy with Fiber Coupled Grazing Angle Probe/FTIR," *Sens. Imaging,* 10 (1): 1-13.

[28] R. Saito, *Physical Properties of Carbon Nanotubes*, Imperial College Press, London, 1998.

[29] Huafeng Yang, Changsheng Shan, Fenghua Li, Dongxue Han, Qixian Zhang and Li Niu. (2009) "Covalent functionalization of polydisperse chemically-converted graphene sheets with amine-terminated ionic liquid" *Chem. Commun.* 3880–2.

[30] S. Guo, D. Wen, Y. Zhai, S. Dong, E. Wang (2011) "Ionic liquid–graphene hybrid nanosheets as an enhanced material for electrochemical determination of trinitrotoluene" *Biosens. Bioelectron.* 26:3475–81.

[31] Joseph Wang, Samo B. Hocevar, Bozidar Ogorevc. (2004) "Carbon nanotube-modified glassy carbon electrode for adsorptive stripping voltammetric detection of ultratrace levels of 2,4,6-trinitrotoluene" *Electrochem. Commun.* 6:176–9.

[32] Po-Chiang Chen, Saowalak Sukcharoenchoke, Koungmin Ryu, Lewis Gomez de Arco Alexander Badmaev Chuan Wang Chongwu Zhou, (2010) "2,4,6-Trinitrotoluene (TNT) Chemical Sensing Based on Aligned Single-Walled Carbon Nanotubes and ZnO Nanowires" *Adv. Mater.* 22:1900–1904.

[33] Park, M., Cella, L. N., Chen, W., Myung, N. V., & Mulchandani, A. (2010). "Carbon nanotubes-based chemiresistive immunosensor for small molecules: detection of nitroaromatic explosives." *Biosensors & bioelectronics*, 26(4):1297-301.

[34] Teresa Demeritte, Rajashekhar Kanchanapally, Zhen Fan, Anant Kumar Singh, Dulal Senapati, Madan Dubey, Eugene Zakar, Paresh

Chandra Ray, (2012) "Highly efficient SERS substrate for direct detection of explosive TNT using popcorn-shaped gold nanoparticle-functionalized SWCNT hybrid," *Analyst*, 137:5041.

[35] Haibo Zhou, Zhongping Zhang, Changlong Jiang, Guijian Guan, Kui Zhang, Qingsong Mei, Renyong Liu, Suhua Wang (2011) "Trinitrotoluene Explosive Lights up Ultrahigh Raman Scattering of Nonresonant Molecule on a Top-Closed Silver Nanotube Array," *Anal. Chem.*, 83:6913–7.

[36] Adem Yildirim, Handan Acar, Turan S. Erkal, Mehmet Bayindir, Mustafa O. Guler (2011) "Template-Directed Synthesis of Silica Nanotubes for Explosive Detection," *ACS Appl. Mater. Interfaces*, 3:4159–4164.

[37] Tae Hyun Kim, Byung Yang Lee, Justyn Jaworski, Keisuke Yokoyama, Woo-Jae Chung, Eddie Wang, Seunghun Hong, Arun Majumdar, Seung-Wuk Lee, *Selective and Sensitive TNT Sensors Using Biomimetic Polydiacetylene - Coated CNT-FETs*, doi: 10.1021/nn103324p.

[38] Fernando Patolsky, Gengfeng Zheng, and Charles M. Lieber (2006) "Nanowire-Based Biosensors." *Anal. Chem.* 78:4260-9.

[39] Eric Stern, Aleksandar Vacic, Mark A. Reed. (2008) "Semiconducting Nanowire Field-Effect Transistor Biomolecular Sensors" *IEEE Trans. Electron Devices* 55(11):3119-30.

[40] Yi Cui, Qingqiao Wei, Hongkun Park, Charles M. Lieber. (2001) "Nanowire Nanosensors for Highly Sensitive and Selective Detection of Biological and Chemical Species" *Science*, 293:1289-92.

[41] Fernando Patolsky, Gengfeng Zheng, Oliver Hayden, Melike Lakadamyali, Xiaowei Zhuang, and Charles M. Lieber. (2004) "Electrical detection of single viruses." *Proc. Natl. Acad. Sci. USA.* 101(39):14017-22.

[42] Gengfeng Zheng, Fernando Patolsky, Yi Cui, Wayne U Wang and Charles M Lieber, (2005) "Multiplexed electrical detection of cancer markers with nanowire sensor arrays." *Nat. Biotechnol.*, 23, 1294-1301.

[43] Brian P. Timko, Tzahi Cohen-Karni, Guihua Yu, Quan Qing, Bozhi Tian and Charles M. Lieber, (2009) "Electrical Recording from Hearts with Flexible Nanowire Device Arrays" *Nano Lett.* 9(2), 914-8.

[44] Z. Li, Y. Chen, X. Li, T. I. Kamins, K. Nauka, R. S. Williams. (2004) "Sequence-Specific Label-Free DNA Sensors Based on Silicon Nanowires" *Nano Lett.* 4(2), 245-7.

[45] Yoni Engel, Roey Elnathan, Alexander Pevzner, Guy Davidi, Eli Flaxer, and Fernando Patolsky, (2010) "Supersensitive Detection of Explosives by Silicon Nanowire Arrays," *Angew. Chem. Int. Ed.* 49, 6830 –5.

[46] Danling Wang, Antao Chen, Alex K. Y. Jen, (2013) "Reducing cross-sensitivity of TiO_2-(B) nanowires to humidity using ultraviolet illumination for trace explosive detection," *Phys. Chem. Chem. Phys.*, 15, 5017.

[47] Sabahudin Hrapovic, Ehsan Majid, Yali Liu, Keith Male, and John H. T. Luong, (2006) "Metallic Nanoparticle-Carbon Nanotube Composites for Electrochemical Determination of Explosive Nitroaromatic Compounds," *Anal. Chem.* 78, 5504-12.

[48] Renyong Tu, Bianhua Liu, Zhenyang Wang, Daming Gao, Feng Wang, Qunling Fang, Zhongping Zhang. (2008) Amine-Capped ZnS-Mn^{2+} Nanocrystals for Fluorescence Detection of Trace TNT Explosive, *Analytical Chemistry*, 80:3458-65.

[49] Chenggen Xie, Bianhua Liu, Zhenyang Wang, Daming Gao, Guijian Guan, and Zhongping Zhang, (2008) Molecular Imprinting at Walls of Silica Nanotubes for TNT Recognition, *Analytical Chemistry*, 80(2):437-43.

[50] Jian Yang, Sharraya Aschemeyer, H. Paul Martinez and William C. Trogler, (2010) Hollow silica nanospheres containing a silafluorene–fluorene conjugated polymer for aqueous TNT and RDX detection, *Chem. Commun.*, 46, 6804–6.

[51] Maung Kyaw Khaing Oo, Chia-Fang Chang, Yuze Sun and Xudong Fan. (2011) Rapid, sensitive DNT vapor detection with UV-assisted

photo-chemically synthesized gold nanoparticle SERS substrates, *Analyst,* 136:2811-7.

[52] H. Ko, V. V. Tsukruk, (2008) "Nanoparticle-Decorated Nanocanals for Surface-Enhanced Raman Scattering," *Small,* 4: 1980-4.

[53] Yingxin Ma, Hao Li, Shan Peng, and Leyu Wang, (2012) "Highly Selective and Sensitive Fluorescent Paper Sensor for Nitroaromatic Explosive Detection," *Anal. Chem.* 84:8415−21.

[54] Subarna Banerjee, Susanta K Mohapatra, ManoMisra, Indu B Mishra, (2009) "The detection of improvised non-military peroxide based explosives using a titania nanotube array sensor," *Nanotechnology* 20:075502-8.

[55] Igor N. Germanenko, Shoutian Li, M. Samy El-Shall. (2001) "Decay Dynamics and Quenching of Photoluminescence from Silicon Nanocrystals by Aromatic Nitro Compounds," *J. Phys. Chem. B* 105:59-66.

[56] Muhammet Erkan Kose, Barbara A. Harruff, Yi Lin, L. Monica Veca, Fushen Lu, Ya-Ping Sun. (2006) "Efficient Quenching of Photoluminescence from Functionalized Single-Walled Carbon Nanotubes by Nitroaromatic Molecules," *J. Phys. Chem. B,* 110: 14032-4.

[57] Yufei Yang, Huaming Wang, Kai Su, Yuanyuan Long, Zhou Peng, Na Li and Feng Liu, (2011) "A facile and sensitive fluorescent sensor using electrospun nanofibrous film for nitroaromatic explosive detection," *J. Mater. Chem.,* 21:11895-900.

[58] Yanke Che, Xiaomei Yang, Guilin Liu, Chun Yu, Hongwei Ji, Jianmin Zuo, Jincai Zhao, Ling Zang, (2010) "Ultrathin n-Type Organic Nanoribbons with High Photoconductivity and Application in Optoelectronic Vapor Sensing of Explosives," *J. Am. Chem. Soc.,* 132, 5743–50.

[59] Yolanda Salinas, Estela Climent, Ramon Martinez-Manez, Felix Sancenon, M. Dolores Marcos, Juan Soto, Ana M. Costero, Salvador Gil, Margarita Parra, Alberto Perez de Diego, (2011) "Highly selective and sensitive chromo-fluorogenic detection of the Tetryl

explosive using functional silica nanoparticles," *Chem. Commun.*, 47:11885–7.

[60] F. A. Calzzani Jr., R. Silesh, A. Kassu, J. M. Taguenang, A. Chowdhury, A. Sharma, P.B. Ruffin, C. Brantley, E. Edwards.(2008) "Detection of residual traces of explosives by surface enhanced raman scattering using gold coated substrates produced by nanospheres imprint technique," *Proceedings of SPIE: Optics and Photonics in Global Homeland Security IV*, Orlando, FL, 9451O.

[61] Nahla A. Hatab Gyula Eres Paul B. Hatzinger Baohua Gu. (2010) "Detection and analysis of cyclotrimethylenetrinitramine (RDX) in environmental samples by surface-enhanced Raman spectroscopy" *J. Raman Spectrosc.* 41:1131-6.

[62] Michael Riskin, Ran Tel-Vered, Itamar Willner. (2010) "Imprinted Au-Nanoparticle Composites for the Ultrasensitive Surface Plasmon Resonance Detection of Hexahydro-1,3,5-trinitro-1,3,5-triazine (RDX)." *Adv. Mater.* 22:1387–91.

[63] Rasmus Schulte-Ladbeck, Martin Vogel, Uwe Karst. (2006) "Recent methods for the determination of peroxide-based explosives" *Anal. Bioanal. Chem.* 386:559–65.

[64] Subarna Banerjee, Susanta K. Mohapatra, Mano Misra and Indu B Mishra. (2009) "The detection of improvised nonmilitary peroxide based explosives using a titania nanotube array sensor" *Nanotechnology* 20(07)5:502.

[65] Sabahudin Hrapovic, Ehsan Majid, Yali Liu, Keith Male, and John H. T. Luong (2006) "Metallic Nanoparticle−Carbon Nanotube Composites for Electrochemical Determination of Explosive Nitroaromatic Compounds." *Anal. Chem.* 78:5504–12.

[66] Sehoon Chang, Hyunhyub Ko, Srikanth Singamaneni, Ray Gunawidjaja and Vladimir V. Tsukruk (2009) "Nanoporous Membranes with Mixed Nanoclusters for Raman-Based Label-Free Monitoring of Peroxide Compounds" *Anal. Chem.* 81:5740–8.

In: Introduction of Forensic Nanotechnology ... ISBN: 978-1-53616-040-6
Editors: R. K. Shukla and A. Pandya © 2019 Nova Science Publishers, Inc.

Chapter 6

EXPLOITING THE POTENTIAL OF DNA FINGERPRINTING IN FORENSIC SCIENCE

Gargi Bhattacharjee*, Khushal Khambhati*,
Nisarg Gohil, Happy Panchasara, Shreya Patel
and Vijai Singh†
School of Biological Sciences and Biotechnology,
Institute of Advanced Research, Gandhinagar, Gujarat, India

ABSTRACT

DNA fingerprinting is a revolutionized technique for crime investigation, individual and relationship analysis, inherited disorders and migratory pattern analysis. Based on the repetitive sequences within DNA, the DNA-fingerprinting techniques have emerged as the dark horse though the complete exploitation of knowledge still remains to be unveiled. The repetitive sequences have been found to render uniqueness to the genome. These distinct DNA sequences can be identified even from a strand of hair or a drop of blood or even flakes of skin. It is currently a key forensic technology with a wide range of applications. In the present chapter, we explore the main concepts of DNA fingerprinting,

---

*Both authors have contributed equally.
†Corresponding Author's E-mail: vijaisingh15@gmail.com.

different techniques that are commonly used in forensic laboratories including RFLP, RAPD, STR, AFLP, mtDNA and touch DNA. Furthermore, nanotechnological application for enhancing the specificity of the PCR, consolidating quantity and quality of the DNA, rapid DNA extraction, constructing DNA biosensors and lab-on-chips have also been mentioned. The chapter also throws light on ethical dilemmas, strength and limitation of the listed techniques.

Keywords: DNA, Forensic Science, DNA fingerprinting, RAPD, STR, RFLP, PCR, AFLP, mtDNA analysis, Touch DNA, Nanoforensic

1. Introduction

Deoxyribonucleic acid (DNA) which is regarded as the ringmaster of life, is an indispensable element of the cellular system. It is a molecule that holds all the consequential instructions required for survival, growth, and reproduction of the biotic community. Following the discovery of the double helix structure of the DNA by Watson and Crick in the middle of the 19th century, scientists have developed an understanding for curing inherited diseases (Růžička et al., 2017). They have deciphered the triplet code of amino acids (Koonin and Novozhilov, 2009) to explore genetic engineering (Resnik and Vorhaus, 2006) and DNA fingerprinting techniques (Jeffreys and Pena, 1993; Sullivan, 1994; Roewer, 2013); and currently it is extensively used in genome editing techniques for creating a robust microbial strain as well as plant and animal breeds (Carroll, 2016, 2017; Singh et al., 2017, 2018). The blueprint of every cell is made up of codes that contain commands to make the essential proteins and molecules in the form of nitrogenous bases: adenine (A), guanine (G), cytosine (C), and thymine (T).

Forensic science is a field that focuses on identification and evaluation of pieces of evidence that are collected from the crime scene. It involves the collection and assessment of fingerprints, shoe-print, handwriting, tire marks, bullets and ballistics, bite marks, scars and biological samples such as blood, hair, bone etc (Saks and Koehler, 2005). According to data concerning 86 cases of DNA exoneration provided by the Innocence

Project, Cardozo School of Law (New York, NY), forensic science testing errors are the second most (63%) contributing factor to wrongful judgment (Saks and Koehler, 2005). In the case of defragmented, decomposed or burnt bodies, forensic identification becomes very difficult. In order to overcome such situations, DNA analysis plays a crucial role because in human DNA sequences, approximately 0.1% sequence is unique to each individual (Collins et al., 1997) and that can be used for investigation purpose.

According to Human Genome Project (HGP), the world's largest collaborative research programme for sequencing the human genome, in human genome only ~1.5% genome encodes a protein and remaining DNA is referred as "junk DNA" (Lander, 2011). A significant portion of the junk DNA consists of specific repeating sequences that are known as variable number tandem repeats (VNTRs) or minisatellites. In 1984, a British geneticist, Alec Jeffreys used these distinctive tandem repeats to build a unique inheritable pattern called "DNA fingerprints" and the technique is well-known as DNA fingerprinting. DNA fingerprinting is one of the greatest revolutionized discoveries of the 20th century (Roewer, 2013) that could be used to distinguish individuals of the same species even from a tiny amount of biological sample. It has been used predominantly in the forensic science for identification of criminal suspects by using biological samples or for recognizing damaged dead body and also for medical uses, for instance, detecting hereditary diseases, finding suitable tissue donor for organ transplantation, linking kinship and blood relatives and so on (Sullivan, 1994; Saad, 2005). In this chapter, we highlight and discuss the different types of DNA fingerprinting techniques in forensic science applications.

2. Emergence of DNA Fingerprinting

The science of DNA fingerprinting and its application in forensic science dates back to late 1980s and was pioneered by Dr Alec Jefferys. A geneticist from the University of Leicester, Dr Jefferys in those days was working on methods to resolve paternity disputes and migration patterns, and was focusing on devising ways to link molecular biology with genetic build-ups. After scrupulous analysis of DNA of different individuals, he deduced that VNTR are short repetitive fragments of DNA which are present in every human being and its length varies in each and every individual unless the two persons are monozygotic or identical twins. It was soon realized that these unique patterns in a person's genome can be employed in the identification of an individual through a technique called DNA fingerprinting (Saad, 2005).

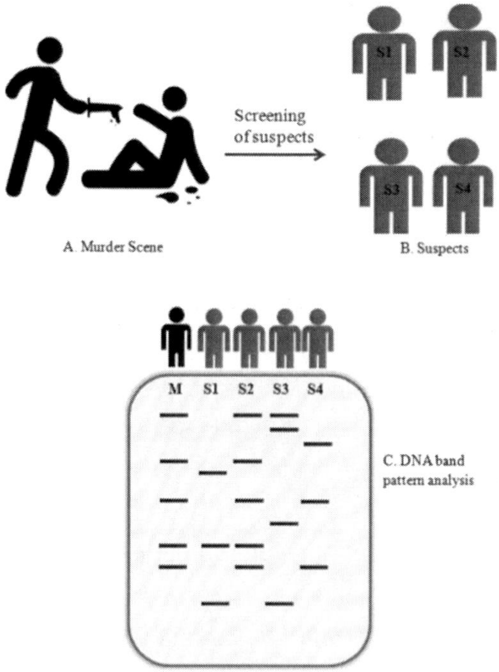

Figure 1 Schematic of Dr Alec Jeffreys DNA fingerprinting analysis. "A" represents a crime-scene where a person is seen to be murdered. Evidences are collected from the spot and are sent for DNA analysis. "B" shows the probable suspects, S1, S2, S3 and S4 namely, who may be involved in the murder. Screening of the evidences and DNA samples has been depicted in "C". Amplification of the VNTR fragments of the suspects reveal a pattern unique to each and every individual. The suspect whose uniquely inherited VNTR pattern matches to the DNA sample derived from the crime scene is convicted of the crime. In figure C, lane M shows the DNA band pattern of the suspected murderer. Since the pattern matches with DNA band pattern of suspect 2 (S2), therefore the person is considered guilty.

Jefferys's DNA fingerprinting technique for the first time ever was used to convict a guilty person involved in two murder cases. In the small town of Leicestershire, Birmingham, two young girls were found raped and murdered. DNA fingerprinting based on restriction fragment length polymorphism (RFLP) of DNA from the semen samples of 4582 men from three nearby towns was performed. VNTR patterns revealed that a local baker, Colin Pitchfork was guilty of the heinous crime of raping and killing both girls (Saad, 2005; Roewer, 2013; Jobling, 2013). In 1987, Colin Pitchfork became the first person in the world to be identified and convicted based on DNA evidence and was sentenced imprisonment for 30 years (Saad, 2005). In 1994, Prof. Jefferys was knighted by the Queen of England for his guardianship and contribution to DNA fingerprinting (Dubey, 2014). Figure 1 shows how DNA fingerprinting technique has been developed and has worked for identification and analysis of samples at the molecular level. This technique is widely used and accepted by the court for better judgment. Basically, the samples from crime-spot are collected and DNA is isolated. It is further used for the processing of samples by PCR and restriction digestion. These samples are then loaded onto an agarose gel in order to perform electrophoresis. Once the samples have run properly the gel-image is analyzed to interpret the band patterns. The DNA fingerprinting results may help to determine biological parents of the child or may be used to figure out a criminal out of suspects.

In India, Dr Lalji Singh, pioneer of DNA fingerprinting technology in India, established a Centre for DNA Fingerprinting and Diagnostic at Hyderabad. Dr Singh and his team could successfully solve a number of baffling cases in India (Mukherjee, 2013). DNA fingerprinting has been used for identification and analysis of the infamous Rajiv Gandhi assassination case of 1992 (Mukherjee, 2013). It was even used to solve the most lime-lighted Naina Sahni tandoor case. Naina Sahni was killed by her husband Sushil Sharma, a political leader on 2nd July 1995 in Delhi. Sushil Sharma had mercilessly shot three bullets killing Naina after which he took her corpse to his restaurant, chopped into smaller pieces and set fire in a tandoor (a traditional clay made Indian cuisine oven) to destroy the evidence. In this case, the DNA fingerprinting technique was used to confirm the brutal killing and spoliation of evidence (Chawla, 1998; Dubey, 2014). Another prominent case is of the parentage dispute involving the political leader N.D. Tiwari and his son from another marriage, Rohit Shekhar. In 2008, Rohit Shekhar claimed himself to be the biological son of Congress leader N.D. Tiwari and this was proved through DNA fingerprinting (Patel et al., 2013).

3. DNA Fingerprinting Methods

DNA Fingerprinting has been the most trusted type of forensic evidence in the world of forensic science. Since its advent, immense progress has been made in the methods of the DNA fingerprinting. Generally, four steps are performed thoroughly in forensic DNA analysis: (1) Evidence collection: sample is collected and transferred to the forensic laboratory for further analysis; (2) DNA isolation: DNA is extracted from the sample of biological origin to obtain purified DNA using different methods such as organic extraction, ion-exchange, solid-phase extraction, etc.; (3) PCR amplification: specified repetitive regions of DNA are amplified to enhance the amount of DNA for subsequent detection; (4) DNA fingerprinting: to obtain the distinctive profile of the DNA, numerous methods have been used including restriction fragment length polymorphism (RFLP), random amplification of polymorphic DNA (RAPD), short tandem repeats/simple sequence repeats (STR/SSR), amplified fragment length polymorphism (AFLP), mitochondrial DNA (mtDNA) and touch DNA analysis to confirm the relation of the sample with the suspect. The details of these different DNA fingerprinting has been discussed here.

3.1. Restriction Fragment Length Polymorphism (RFLP)

RFLP uses restriction endonuclease in order to achieve successful DNA fingerprints (Botstein et al., 1980). Bacterial cell harbours several types of restriction endonuclease enzymes. These enzymes being a part of the bacterial innate immune system, helps to fight against the invading phages and other extracellular genetic elements (Shabbir et al., 2016). The enzymes recognize particular nucleic acid sequences, which are commonly known as restriction sites. Each and every type of restriction endonuclease recognizes a specific restriction site (Roberts and Murray, 1976). Once the restriction site is recognized by the restriction endonuclease enzyme, it displays its nuclease activity and breaks the phosphodiester bond (a covalent bond in DNA or RNA that joins polynucleotide chains together) (Figure 2).

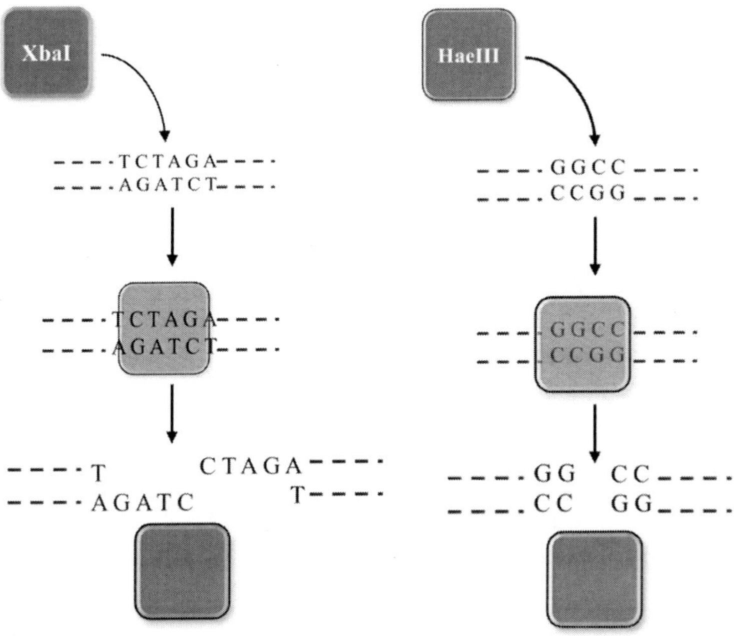

Figure 2. Restriction endonuclease recognizes specific sequences and breaks the phosphodiester bond. Box in blue colour represents XbaI restriction endonuclease producing sticky end cuts while HaeIII (represented in green) gives blunt end cuts

The RFLP technique is based on two facts (Brettschneider, 1998):

1. A particular restriction endonuclease used in RFLP would recognize only a specific site among the nucleic acid sequence and carry out digestion reaction.

2. Each and every individual have a different number of the restriction site, and alteration in one or two base pairs within the restriction site would lead to the formation of a completely different sequence which would not be recognized by the restriction enzyme specific for that particular restriction site sequences.

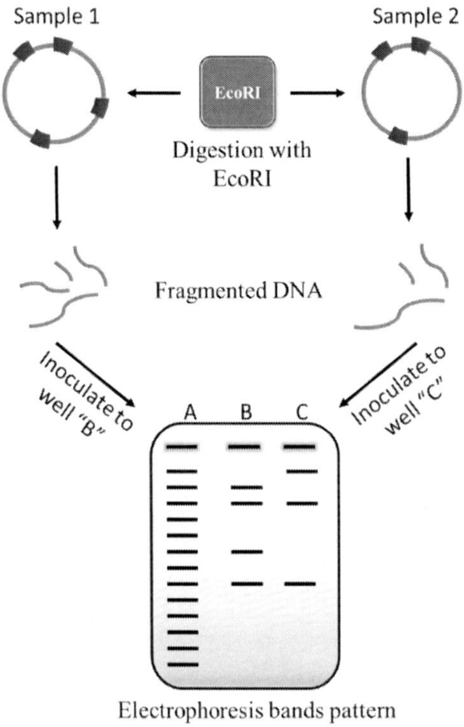

Figure 3 Restriction fragment length polymorphism (RFLP). For simplicity, consider the circles marked with blue as two different genomic DNA. The red colored boxes represent the EcoRI restriction sites. Due to the presence of SNP in genomic DNA of sample 2, it lacks one EcoRI restriction site in comparison to genomic DNA of sample 1. On digestion of both the genomic DNA with EcoRI, varying sizes of DNA fragments would be obtained. The mixture is subjected to electrophoresis and band patterns are detected. Lane 'A' consists of the DNA ladder sequence bands. Lane 'B' consists of bands produced from digested fragments of genomic DNA 1 while lane 'C' consists of band patterns obtained from digestion of the genomic DNA 2 with EcoRI.

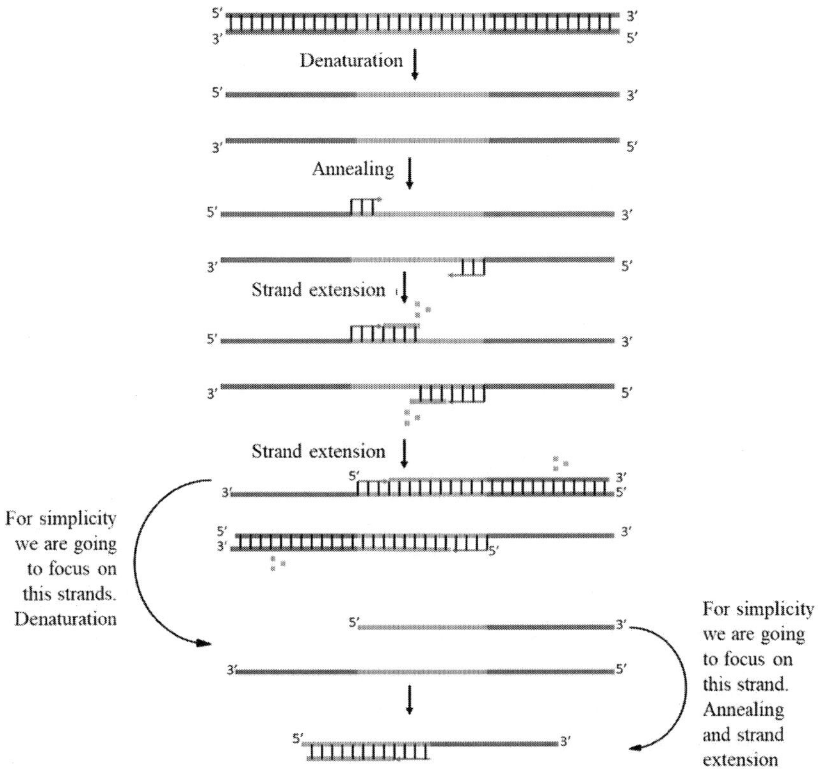

Figure 4. Polymerase chain reaction (PCR). The region marked with green is the desired sequence which is to be amplified from the DNA (red and green region). With initiation of denaturation step, dsDNA are separated into two ssDNA. Annealing at favorable temperature allows the primers (blue arrows) to bind to their complementary sequences on DNA. Lastly, strand extension by thermostable DNA polymerase facilitates extension of the polynucleotide chain by stitching together the incoming deoxynucleotides with the free hydroxyl groups of the terminal nucleotide of the primer via means of a phosphodiester bond. The cycle of denaturation, annealing and strand extension repeats till abundant copies of the desire sequence accumulates.

The RFLP technique treats the isolated genomic DNA with one or two restriction endonuclease enzymes. The resultant reaction would give varying lengths of DNA fragments (Botstein et al., 1980). These DNA fragments are separated by agarose gel electrophoresis. The separated fragments are then transferred onto a nitrocellulose paper where the fragments undergo hybridization reaction. This reaction incorporates some specific probes that could help to detect bands in a unique pattern (Figure 3). On digestion of the total human DNA, RFLP has been found to be a powerful tool for identification of the variants within specific regions of the genome (Botstein et al., 1980).

Based on the principles of RFLP, a new technique has been designed which is called as polymerase chain reaction-restriction fragment length polymorphism (PCR-RFLP) (Rasmussen, 2012). It has diligently combined the potential of the PCR for DNA segment amplification with the ability of the RFLP to discriminate the variants. PCR is a technique that helps to increase the copy number of the desired sequence (figure 4) (Watson, 2012). Simply put, it is a technique to amplifya given set of sequences. PCR involvesthree major steps, that is denaturation, annealing, and extension. In denaturation, the desired sequence of the double-stranded DNA (dsDNA) molecule which is to be amplified is separated by providing high temperature into two single-stranded DNA (ssDNA) molecules. During annealing, the short oligonucleotides (called primers) bind to both of the 3' end of the desired sequences from the ssDNA molecule. Lastly, in the extension step, the polymerization reaction carried out using thermostable DNA polymerase enzyme (Taq DNA polymerase) from the 3' end of the attached primer, allows addition and linking of the incoming deoxynucleotides to the growing polynucleotide chain following strand extension by Watson and Crick base pairing rule (Watson, 2012). These cycles are repeated several times for increasing the copy number of the desired sequence.

In PCR-RFLP, specific DNA segments including the internal transcribed spacer (ITS) and the intergenic spacer (IGS) of the ribosomal nuclear DNA are amplified, which are known to possess possible polymorphs (Buscot et al., 1996) (figure 5). The resultant pools of amplified segments thus obtained are subjected to digestion by a restriction endonuclease. The digested fragments are separated through electrophoresis followed by band detection. One of the advantages of PCR-RFLP over usual RFLP is that it can help to carry out operations at a detectable range even with a very low concentration. Whereas, a major disadvantage of using PCR-RFLP is the difficulty in recognizing the restriction endonuclease sites that would be affected by several SNPs (Rasmussen, 2012).

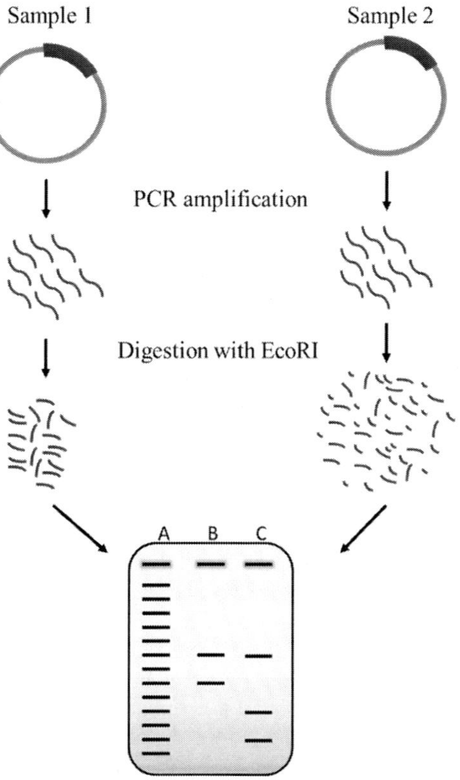

Figure 5. Polymerase chain reaction-Restriction fragment length polymorphism (PCR-RFLP). For simplicity, consider the circles marked with blue as two different genomic DNA. The red colored box represents the sequence that needs to be amplified. In PCR-RFLP the ITS or IGS regions of the ribosomal DNA sequences are amplified. The amplified product obtained as a result of the PCR amplification are subjected to digestion with specific restriction endonuclease enzyme. Here, the genomic DNA of sample 1 contains one restriction site for EcoRI in its genomic IGS or ITS region of the ribosomal DNA sequence. While, the genomic DNA of sample 2 contains two restriction endonuclease sites for EcoRI. On digestion of the amplified segments obtained from the genomic DNA of sample 1 and 2 different sized fragments are generated . Electrophoresis then separates different strands according to their sizes and a unique band pattern is obtained. Lane 'A' shows the DNA ladder bands, while lane 'B' and 'C' depicts the patterns generated from genomic DNA of sample 1 and 2 respectively.

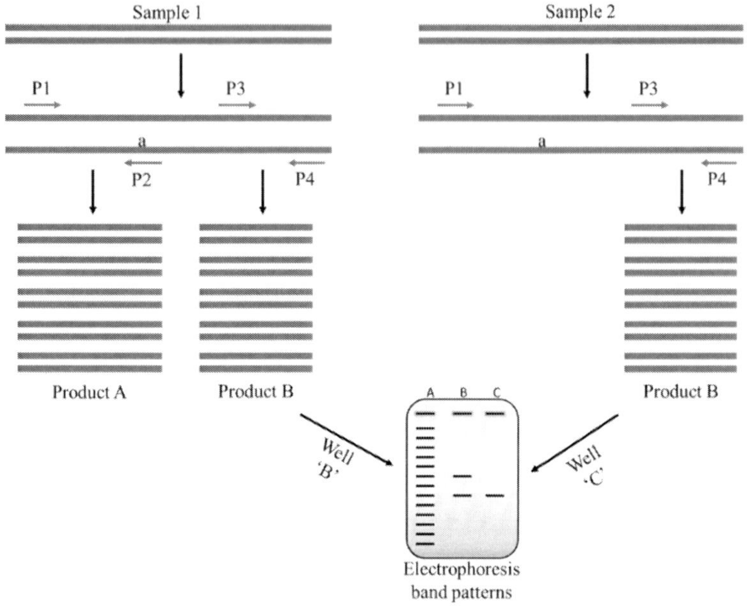

Figure 6. Random amplification of polymorphic DNA (RAPD). Sample 1 and sample 2 are two isolated genomic DNA. During PCR, primers of 10-15 nucleotides of random sequence are used to amplify the sequence in the genome. The primers are denoted as P1, P2, P3 and P4. These primers would find its complimentary region in the genomic DNA and carry out PCR reactions. Sample 2 here, has SNP at loci 'a'. 'a' is a complimentary region where primers would bind. On PCR amplification, sample 1 would give two products (product A and product B) of different lengths. While sample 2 would give only one product "B" because locus 'a' has SNP so the primers cannot bind to that region. Further in sample 2, primer P1 and P4 cannot give PCR product as they are distantly apart. The band pattern of sample 1 genomic DNA is represented in lane B while of sample 2 genomic DNA is represented in lane C. Lane A depict bands of DNA ladder sequence.

3.2 Random Amplification of Polymorphic DNA (RAPD)

Polymorphism is the genetic variation that occurs at a particular locus in the DNA. RAPD is a technique that helps to detect polymorphism by using PCR and the resultant product is subjected to detection of the band patterns (Lynch and Milligan, 1994). Unlike PCR, RAPD uses arbitrary primers in its reaction. In RAPD, primers of random nucleotide are used to amplify the sequences. This also means that in RAPD no prior knowledge about the sequence that is going to be amplified is needed (Lynch and Milligan, 1994). For example, a primer of 10-15 nucleotide designed for RAPD fingerprinting of any random sequence would bind to its complementary regions in the genomic DNA molecule. If the 3' end of the primers lie far away from each other,then the intermediate sequences between those two primers won't be amplified during PCR but if the 3' end of the two primers is in close proximity to each other, then the sequence would be amplified (figure 6) (Lynch and Milligan, 1994). In this way, different lengths of the DNA fragments in a large quantity are generated, following which the fragments are separated based on their sizes and the band pattern is detected through agarose gel electrophoresis (Lynch and Milligan, 1994). A disadvantage of RAPD is that it totally depends on PCR based enzymatic reaction (Kumari and Thakur, 2014). Therefore, the reaction conditions including primer annealing temperature, PCR cycles, the concentration of PCR components as well as the template DNA, altogether influences the outcome of this technique (Lorenz, 2012).

3.3. Short Tandem Repeats (STR) or Simple Sequence Repeats (SSR)

Short tandem repeats (STR) or microsatellites are one such form of DNA polymorphism. These repeat sequences are widely found scattered throughout the genome of prokaryotes and eukaryotes including humans. STRs usually occurs as 1-6 bp short repetitive stretch of DNA (Ham et al., 2016), forming series that appears as repetitive, tandemly stitched sequence of nucleotides lengthening up to 100 nucleotides per repeats (Fan and Chu, 2007). STRs are commonly found within the non-coding regions, however, about 8% are found located within the coding region (Ellegren, 2000; Gymrek et al., 2016). On an average, the frequency of occurrence of STR in the human genome is one per 2000 bp (Lander et al., 2001). STR locus is for example designated as, D19S1234; wherein, D stands for DNA, 19 is the chromosome number on which the STR locus is located, S denotes STR, and 1234 is the unique identification number for that particular STR.

STRs can be classified based on either the length of the repeating units or the basis of the repeat structures. Based on the number of repeating nucleotides, STRs can be classified into mono- or single-, di-, tri-, tetra-, penta-, and hexanucleotide repeats. STRs found within the human genome are of the dinucleotide type. On the basis of the repetitive structures, STRs can be classified into simple or perfect repeats that are sequence stretch containing just one kind of repetitive unit, and compound or imperfect repeats, meaning sequences consisting of a different composition of repeats (Urquhart et al., 1994; Bhukya, 2016).

Though STRs are widely distributed throughout the genome of an organism, conventionally these were thought to have absolutely no biological roles at all. Therefore, these were regarded as "junk DNA". However, due to the advances in technique, several hypotheses have evolved that have suggested that STRs might be playing some interesting roles, which are yet to be unveiled. For instance, products of the STRs with glutamine-rich domains located within a protein-coding region have shown to play a functional role in transcriptional regulation in higher organisms (Serpionov et al., 2015). A number of "contingency genes" of some pathogenic bacteria have been found to be residing in the STR sequences. These STRs may lead to frame-shift mutations in some cases causing alterations in protein expression. These proteins though not necessary for the survival of the bacteria can protect them from killing by the human or host immune response (Fan and Chu, 2007).

The knowledge of STRs has been extended to genetic linkage analysis, derivation of specific gene locations, construction of genetic maps (Li et al., 2017), parentage analysis, as well as diagnostics of some diseases. It is also used for studying population genetics and biological diversity at various biological organization levels (Kim and Sappington, 2013), or to re-construct migratory patterns and evolution of species (Abdul-Muneer, 2014). Conventionally, the STRs were analyzed through amplification of target STR sequences by PCR, where the unique sequences flanking the STR served as a primer. The length of the resultant PCR product was then determined using gel electrophoresis that eventually indicated the copy number of targeted STR (Nims et al., 2010). The rapid development in the next-generation sequencing (NGS) technique has helped to simplify the STR profiling for its incorporation into whole genome sequencing (WGS) data (Tang and Nzabarushimana, 2017).

3.4. Amplified Fragment Length Polymorphism (AFLP)

AFLP is a technique which is widely employed in DNA fingerprinting. AFLP utilizes the adapter molecules which are nothing but short synthetic double-stranded DNA molecules with single-stranded flanking ends (figure 7). This flanking region is complementary to specific DNA sequence (Paun and Schönswetter, 2012).

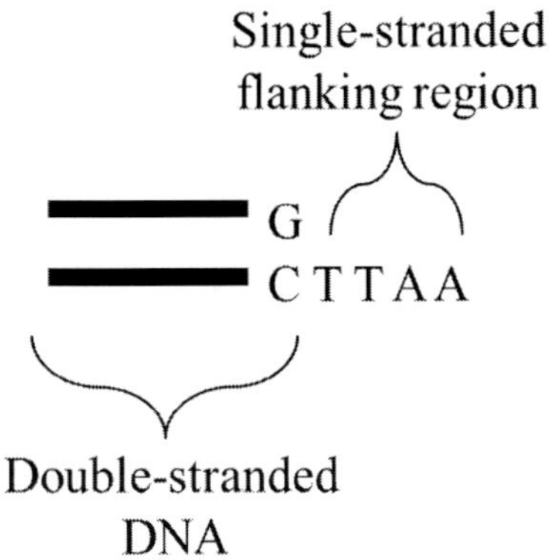

Figure 7. Synthetic adapter molecule

Mechanism of AFLP goes as follow (Vos et al., 1995) (figure 8).

1. The isolated genomic DNA is subjected to usually two restriction enzymes. One would be a rare cutter and other a frequent cutter. The frequent cutter occurs frequently and generates numerous small fragments. On the contrary, the rare cutter enzyme occurs

rarely in a genome (e.g., Not I). The enzymes selected should be able to produce sticky ends.

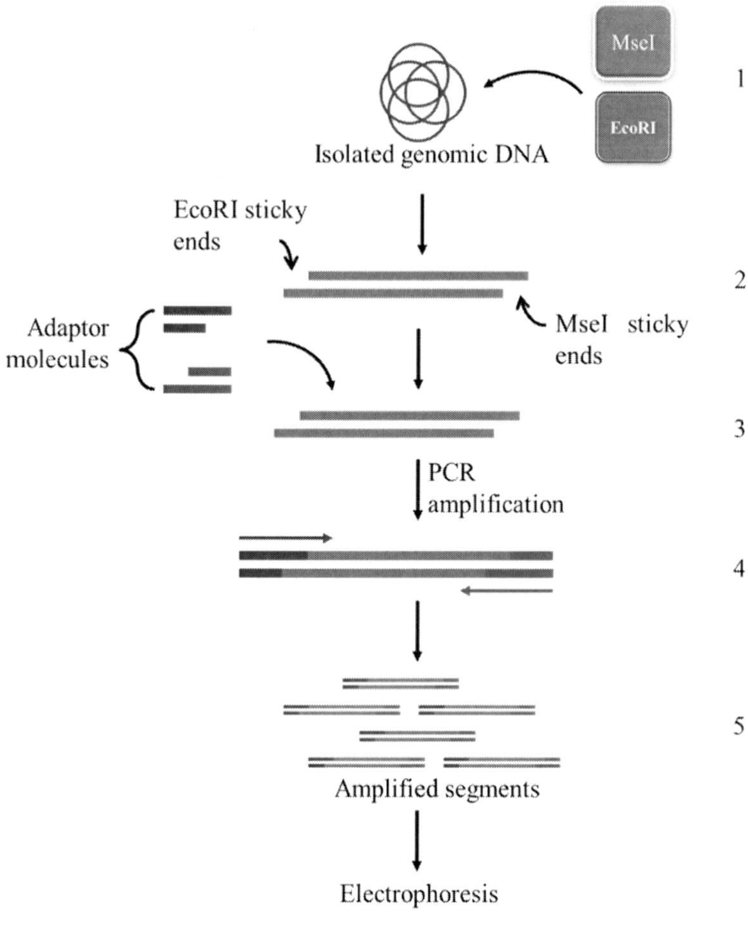

Figure 8. Amplified fragment length polymorphism (AFLP). During AFLP the isolated genomic DNA sequences are digested with two different restriction endonucleases (1), one being frequent cutter and the other being rare cutter. This leads to formation of different lengths of DNA fragments. Some of the fragments harbors rare cutter sticky end on one side while frequent cutter sticky end on the other. One of such fragment is shown in (2). The adaptor molecule complementary to the sticky ends are added and ligated (3). The complementary sequences would be amplified (4) (5). Likewise many other different sized fragments would be amplified simultaneously (not shown here). These fragments are subjected to electrophoresis and the unique band pattern are analyzed.

2. Once the genomic DNA is digested with a restriction endonuclease, the adapter molecules are added and ligated with the aid of ligase enzyme. The adapter molecules are synthesized in a way that its flanking end would be complementary to the sticky ends that are produced by restriction endonuclease on digestion of the isolated DNA. The resultant sequences formed is subjected to PCR.
3. The PCR product is run on agarose gel electrophoresis for separating the amplified bands. The labelled primers used during electrophoresis would aid in highly sensitive band detection.

The 3' end of the forward and reverse primers used during the DNA amplification, which is flanked by 2-3 random nucleotides, greatly restricts its possibilities to attach to non-specific sequences (Vos et al., 1995).

3.5. Y-Chromosomal Short Tandem Repeats (Y-STR) Analysis

Y-chromosomal short tandem repeats (Y-STR) polymorphism is a method to characterize male-specific part of the human Y-chromosome or for inferring biological sex of the contributor (von Wurmb-Schwark et al., 2003; Kayser, 2017). The Y-STRs show a great deal of diversity amongst the males of unrelated origin and the diversity more or less remains the same over the generation. The Y-STR haplotyping proves to be advantageous particularly in cases where the standard autosomal DNA profiling seems uninformative (Kayser, 2017). Y-STR haplotypes may help (i) include or exclude the probable male suspects involved in a crime, (ii) to trace multiple male donors, (iii) provide lead to detect unknown male offenders and (iv) to draw paternal lineage (Kayser, 2017).

In case of sexual assault, the DNA sample recovered from vaginal swabs will contain DNA sample from both, the female as well as the male culprit(s). Though differential extraction attempts to separate the DNA of both sexes, it is impossible to achieve error-free separation completely. In such circumstances, the female DNA gets simultaneously amplified along with the amplification of the male DNA, which may mask the DNA profile of male making the analysis difficult. To circumvent the masking errors, Y-STRs are examined (Kayser, 2017). Since the DNA from the female by no means can contribute to the Y-STR, the male suspects can be easily detected and the variation in the Y-STR pattern can clearly reveal the possible male culprits. Other than this, the knowledge of Y-STR analysis can be extended to non-sexual assault cases such as to determine the sex (Aal-Hamdan et al., 2015) and for identification of disaster victims or unclaimed bodies (Ambers et al., 2018) and to solve paternity disputes between the male offspring and related kinships (Rolf et al., 2001).

3.6. Mitochondrial DNA (mtDNA) Analysis

Sometimes, a sample fetched for DNA analysis is found to be damaged, old, degraded, or quite less to be analyzed by the RFLP and STR analysis (Budowle et al., 2003). In such cases, mitochondrion organelle may play a pivotal role to solve the mystery. Mitochondria, the powerhouse of the cell, have extranuclear, histone-free, double-stranded, circular, distinct, high copy number DNA. Owing to its higher intracellular concentration as compared to nuclear DNA, maternal inheritance, and its lower susceptibility to undergo degradation, mtDNA is often employed in low template DNA detection (Budowle et al., 2003). Due to their highly distinctive polymorphic characteristics, the hypervariable region, especially regions I and II (HVI and HVII) of the mtDNA are used for analyzing utterly degraded DNA samples (Budowle et al., 1999). It could be used for DNA fingerprinting for identification of a person.

3.7. Touch DNA Analysis

Human beings shed roughly 4,00,000 skin cells every day (Wickenheiser, 2002) and these cells may transfer to every single object, we touched and handled. When a crime is committed, the body fluids or epithelial cells left behind unintentionally by the perpetrator on the crime scene can be used to crack the identity of the perpetrator. The touch DNA is a tool for rapid collection of samples (Williamson, 2012). It just requires a minute amount of skin cells to solve the crime mystery.

Table 1. Merits and demerits of DNA analysis techniques in a nut-shell

DNA analysis	Merits	Demerits
SNP	Single nucleotide variations can be detected even from the highly degraded or old DNA sample (Butler et al., 2007).	Less polymorphic, cannot be used for detecting DNA mixture and is relatively less informative per locus compared to STR (Butler et al., 2007).
STR	Most precise analytical method for evaluating DNA from soft tissues or remains of bone till now (Nims et al., 2010).	Cannot be used for studying low-template or extremely poor quality of DNA samples (Nims et al., 2010).
Y-STR	Capable of specifically detecting and distinguishing the male DNA specimen from a DNA mixture pool containing female DNA (Printz, 2003).	Male-specific DNA analysis. Inability to uniquely identify male relatives (Printz, 2003).
mtDNA	mtDNA being present in higher proportion than nuclear DNA and is less prone to undergo degradation, is favoured to be employed for highly degraded and low-template DNA analysis (Kline et al., 2005).	Extraction is tedious and expensive. Inheritance from mother or father cannot be precisely predicted (Strehle et al., 2018).
DNA Methylation	Biochemical addition of methyl groups to adenine and cytosine aids better and specific identification of body fluids, concurrent with STR protocol (Vidaki et al., 2013).	Demands larger sample size, is laborious and tedious (Vidaki et al., 2013).
Microfluidic Systems and Nanoforensic	Both microdevices and incorporation of nanoparticles greatly reduces the required sample size and reagent volume, enhances specificity of the PCR and also strengthens quality and quantity of DNA (Bruijns et al., 2016).	Often fails to quantify DNA (Benítez et al., 2012)

The trace amounted DNA sample is employed for isolation and PCR amplification and subsequently used for building highly specific genetic portrait through length polymorphism and sequence polymorphism (Sowmyya, 2016). However, the technique comes with several limitations such as DNA contaminations, difficulties in elucidating trace DNA mixtures and financial constraints (Sowmyya, 2016). Nevertheless, it has helped to resolve many cold cases where the recognition of the culprit was a challenging task for forensic experts (Sowmyya, 2016). The merits and demerits of principle DNA fingerprinting techniques have been shown in Table 1.

4. Forensic DNA Mixture Analysis

In forensic analysis, the DNA samples are often recovered as a mixture of multiple DNAs originating from more than one individual. The principal objective in such cases is to clarify or confirm individual DNA profiles from a mixed DNA sample and also determine the mixture ratios. Currently, STR polymorphism analysis and assessment of mtDNA variations are commonly employed techniques in forensic investigation of DNA mixtures (Holt et al., 2002) and the credit for its ease and efficiency goes to the up-gradation in DNA sequencing. With the obvious improvements in the DNA extraction methods and some manipulations in the PCR cycles, it is now possible to detect even extremely poor, low-template and highly complex DNA mixtures (Bieber et al, 2016). Following amplification of the STRs at distinct genetic loci, the resultant amplified products are separated according to their sizes through means of capillary electrophoresis (CE). The presence of more than two allelic peaks at different genetic loci or the difference in the peak heights in comparison to a heterozygotic peak height ratio is a clear-cut indication of the presence of DNA contributed by multiple donors (Bieber et al., 2016).

At times, very low amount of DNA on amplification tends to show stochastic effects; wherein either one or both the allelic copies of the STR at a given locus fail to get amplified, thus, resulting into an infrequency known as allelic dropout (Gill et al., 2000). In such cases, where the genomic data is stochastic, only statistical prediction of the data remains a choice. To deal with such dilemmas, forensic experts across the globe rely on combined probability of inclusion (CPI) (Ladd et al., 2001). CPI analyses the probability of the portion of population that can be possibly included in the list of potential contributors to a given DNA mixture. Interpretation of data through the CPI involves three basic steps; first is the analysis or evaluation of the retrieved DNA mixture profile, which is followed by comparison of the data with the reference profiles in order to include or exclude probable donors. Lastly, the data is statistically interpreted to be presented at the courtrooms. The counterpart of CPI is combined probability of exclusion (CPE).

5. DNA PHENOTYPING

The biological evidences are mostly screened for the serological variations (Gefrides and Welch, 2006; Peterson et al., 2010), per se, ABO blood typing (Wynbrandt and Chisum, 1971; Ramnarayan et al., 2013; Khairkar et al., 2015) or isoenzyme polymorphisms (Sensabaugh, 1982) in order to screen out the possible number of potential culprits. A serious setback of the entire process is that the serology aided elimination could only reduce the donor pool but individualizing a particular donor was nearly impossible (Gefrides and Welch, 2006). With rapidity in the technological advancements, it is now possible for analysts to develop a physical profile with typically unique and distinctive characteristics. This is known as forensic DNA phenotyping (FDP) for externally visible characteristics (EVCs), wherein, the technique utilizes evidentiary DNA specimen that may have been accidentally left behind by the perpetrator (Kayser, 2015). FDP is applied to analyse the DNA samples from unknown individuals, especially, when it is difficult to eliminate a potential suspect by any other investigative means. Once the list of suspects is narrowed down to a few potential suspects using FDP assisted EVCs, subsequently conventional DNA techniques can be applied to confirm the identity of the individual (Kayser and Schneider, 2009).

The underlying principle of FDP is that DNA being the master manipulator of a person's physical and genetic makeup, its knowledge can be extended to predict the physical feature, frame or body structure. The human genome holds millions of variations that may differ by just a single nucleotide commonly referred to as single nucleotide polymorphisms (SNPs). SNPs are thought to be responsible for biologically-influenced development of physical features (Van Laan, 2017). The concept of phenotyping relies on the basic understanding that within the genome there occur specific DNA segments that have high correlation with the expression of certain characters and it is unique nucleotides that impart the uniqueness to a person's physical trait.

There are mainly two forms of DNA phenotyping: direct and indirect. Indirect phenotyping shapes or traces the phenotypic characteristics based on the ethnicity or geographical origin of the individual. Direct phenotyping, on the other hand, deeply analyses the genotypic variations to understand predominant physical characteristics (Koops et al., 2006). FDP for EVCs chiefly focuses on analysis of those portions of the DNA that regulates the physical features of a person, and, to meet this specification, an amalgamation of both of the above mentioned forms of DNA phenotyping is employed (MacLean, 2013). To identify and study SNPs coding for certain physical characters, DNA phenotyping makes use of the PCR-STR technique.

6. Nanotechnology for DNA Analysis

Nanotechnology is a science dealing with an entity that has the scale dimensions ranging from 1-100 nm. It incorporates nanoparticles (NPs) which are extremely small and loaded with extensive properties such as tiny size, high surface area to volume ratio, better electric and catalytic properties making it preferred choice in biomedical, biotechnological, agricultural, industrial field. Currently, NPs have been used to enhance the specificity of the PCR (Yeow and Wan, 2010), as instances, gold NPs (AuNPs) (Li et al., 2005; Vu et al., 2008; Cai et al., 2010; Vanzha et al., 2016), silver NPs (AgNPs) (Rehman et al., 2015), titanium dioxide NPs (Mahmoud et al., 2017), carbon nanotubes (CNTs) (Zhang et al., 2008; Du et al., 2014) and ammonium salt of oleic acid coated magnetite (Fe_3O_4) NPs (Kambli and Kelkar-Mane, 2016).

The advent of nanotechnology has now delivered newer, faster and convenient methods in context to extract DNA from the human biological samples. For example, a magnetic NPs (a class of magnetic field manipulated NPs) and copper NPs (CuNPs) implemented methods (Saiyed and Ramchand, 2007; Saiyed et al., 2008; Zhou et al., 2013; Lodha et al., 2017). The magnetic NPs employed method has an ability to extract the DNA in only 15 minutes with a much higher yield as compared to conventional DNA extraction methods (Saiyed and Ramchand, 2007). Similarly, CuNPs have also been used for the DNA extraction from the skeletal remains and ancient archaeological samples (Lodha et al., 2017). In addition, NPs based detection has consolidated the quality and quantity of DNA (Paul et al., 2014; Al-Jeboory et al., 2015).

Apart from this, NPs are currently used for constructing DNA biosensors (Cherstvy, 2013; Siddiquee et al., 2014; Abu-Salah et al., 2015; Hajihosseini et al., 2016; Saeed et al., 2017). DNA field-effect device (FED) covered with AuNPs is an ideal example of the same. In this biosensor, DNA sequences are pulled toward the AuNPs on the FED surface, enabling a very low level of DNA detection (Cherstvy, 2013). Using NPs the DNA detection has been achieved up to 5 attomolar (10-18 molar) concentration by employing AuNPs (Zhou et al., 2011).

In DNA fingerprinting technique, a number of steps need to be performed such as isolation of DNA, PCR, and electrophoresis. In order to minimize time and cost, a microfluidics platform has been revolutionized. Recently, in order to integrate the steps that utilize instruments for DNA isolation, PCR amplification and their separation along with analysis, scientists have developed lab-on-chip devices (McCord, 2006). These devices perform all the necessary laboratory steps and functions that are required to get an experimental result. Lab-on-chip devices integrate micro-fabrication with microfluidics in order to achieve the above-mentioned steps. The miniaturization gives the advantage of low sample volume (nanolitre range), rapid analysis and minimizes the labour. The chips available are to the size of millimetres to a few square centimetres (Gupta et al., 2016). These chips are divided into separate compartments for performing individual steps (Figure 9). In near future, the forensic laboratory will have a microfluidics-based chip that may be integrated with highly advanced automated microfluidic devices for high throughput and rapid DNA analysis for identification of suspects.

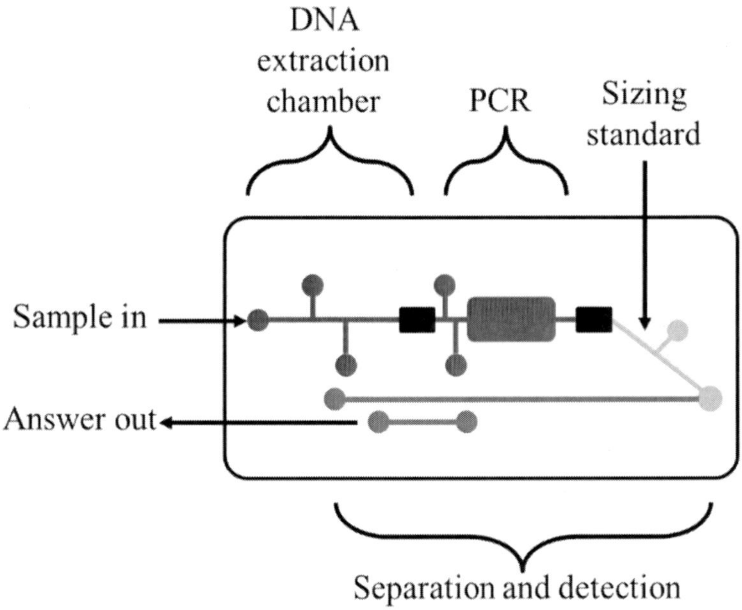

Figure 9. Lab-on-chip. The dots represent reservoirs and wells for input of reaction components as well as removal of waste. Black box represents valve.

7. Potential Applications of DNA Fingerprinting

The novelty and advancement in DNA fingerprinting techniques have been a boon in generating considerable excitement in the criminal justice community. These techniques are now not limited to legal analysis but have also made their way in solving enigma at the molecular level. As shown in Figure 10, the potential applications of DNA fingerprinting in diverse areas for investigations.

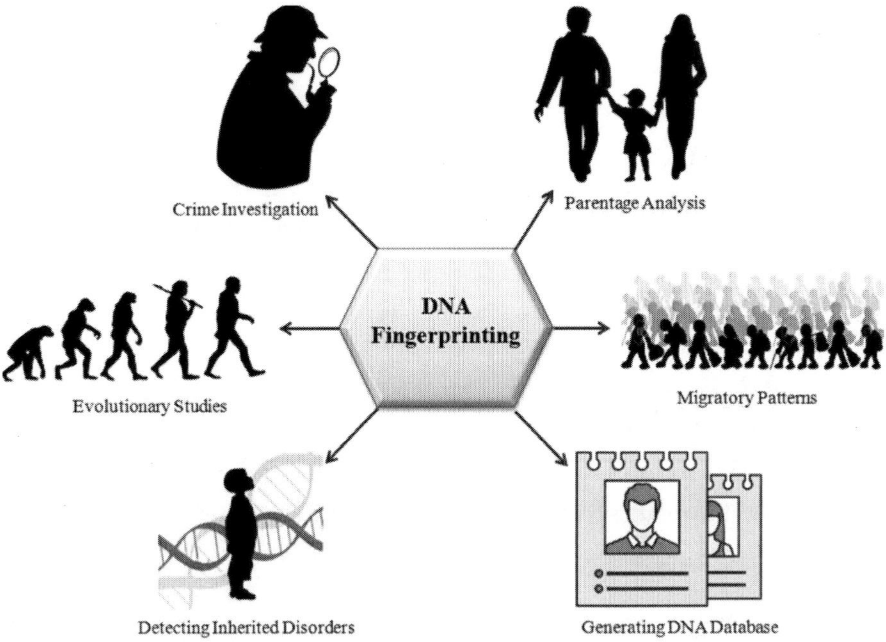

Figure 10. Potential application of DNA Fingerprinting in parentage analysis, determining migratory patterns, generating DNA databases, detecting inherited disorders, evolutionary studies and in investigation of crime.

7.1. Criminal Identification

Courts have widely accepted the DNA fingerprinting techniques as sensitive and accurate proof of case worldwide (Kobilinsky et al., 2005). Samples such as blood, hair, saliva or semen are collected from the crime scene that serves as the decisive evidences for a case. As the DNA in each cell of our body is distinct, one can be identified by the DNA. Even from a minute DNA sample, it is possible to find a guilty from the list of suspects.

A number of cases have been solved globally using the DNA fingerprinting technique (Shrivastava et al., 2012).

7.2. Parentage Testing and Relationship Analysis

The human reproduction involves the formation of a zygote, the first new cell of a child, from the fusion of a sperm and the ovum in a manner that the child shares the DNA from both, the father and mother. Of the 46 chromosomes present in humans, a child inherits 23 chromosomes from the father and remaining 23 chromosomes from the mother. The 22 autosomal pairs of chromosome harbour the same set of genes with very minute differences. These differences in the form of SNPs make each copy of the gene uniquely inherited from father or mother. The 23rd pair of chromosome or the sex chromosome determines the sex of the child. A male bears an X and a Y chromosome while a female bears XX chromosome. During fertilization, if the sperm bearing Y chromosome fuses with the ovum, then the child born will be a male. While if X chromosome is passed in the sperm, the child born will be a female. A mother always passes an X chromosome to the child, but whether the child receives an X chromosome or a Y chromosome from the father ultimately decides the sex of the child. Therefore, the DNA from the child and from its parents can be compared for parentage similarity and analysis. As for relationship analysis, DNA fingerprinting may aid to determine the descendants or heirs to resolve property disputes (Shapiro et al., 1992).

7.3. Identification of Inherited and Genetic Diseases

As the child shares the DNA from both the parents, there is always a possibility that if either parent has an inherited disorder, the trait might pass onto the child. Analyzing faults in DNA patterns obtained from individuals with and without a disorder may predict the forthcoming of genetic disorders. DNA fingerprinting is a rapid process that allows scientist to detect the genetic disorders including haemophilia, thalassemia,sickle cell anaemia and cystic fibrosis using the parents' sample which can pass to children. It could be useful for early diagnosis, prevention and treatment of diseases (Saad, 2005).

7.4. Migratory Pattern Analysis

DNA fingerprinting of a population may help the historians to analyze migratory patterns and to interpret cycles of continuity; epidemiologists to analyze health-related issues in a population which may be attributed to genetic predisposition, and geneticists to study human dispersal patterns to discover and describe genetic variations within a population (Amos et al., 1991; Murray-McIntosh et al., 1998).

This could be tracked through an example of when Sir Alec Jeffreys had to solve a case of a Ghana immigrant family who had immigrated to the UK and had become UK citizens. For some reason, the son of the family had gone to Ghana and while returning to the UK, he wasn't allowed to enter into UK accusing of holding a forged passport. Dr Jeffreys was contacted to solve the case. Samples from the son and others in the family were taken for proving the identity. It was concluded that the individual was the son of the mother and father of UK residents (Jeffreys et al., 1985). This immigration case opened the doors for using DNA fingerprinting to recognize a person's individuality. In 2005, a study published in Science revealed that the knowledge of DNA fingerprinting could be extrapolated for identification and study of the origin of Andaman "Negrito" and Nicobar "Mongoloid" populations. They found that population residing in Andaman shared their common ancestor with a lineage spread among India, East Asia, Africa, New Guinea and Australia (Thangaraj et al., 2005).

7.5. Development of a DNA Database

A pressing need arises to develop a DNA database for all individual for rapidity in resolving legal cases and other issues. US government has already made and synced a DNA database at the local, state and national level which comprises of DNA fingerprints of people with criminal records (CODIS: Combined DNA Index System). It can be now easily shared, exchanged and can be compared with DNA profiles electronically (McEwen, 1995). This was designed based on amplification of 13 core STR loci and amelogenin gene that are present in X and Y chromosome. This database can also help to determine the sex of an individual (Moretti et al., 2001). Currently, every country requires to make an extensive effort for developing such kind of DNA databases to solve the issues of crime, health, and identity.

8. Ethical Concerns

Ethics is the moral values, beliefs or principles that govern the behaviour or conduct of a person. Simply, it is the understanding of what is right and what is not. The foundation of ethical genetic testing lies over the principles of autonomy, confidentiality and privacy.

8.1. Autonomy

The word 'autonomy' is derived from the Greek word *autonomos* (*auto* - "self" and *nomos*- "law"), meaning, actions governed by self-rule or self-determination. Autonomy in the context of genetic testing refers to the right of an individual to independently decide whether the person is willing to go for the test. The conversation regarding the medical samples, autonomy gives the authority to an individual to decide and allow their genetic material or information to be stored and used for future analysis. However, autonomy may be overridden in certain cases such as screening of neonates for hypothyroidism and phenylketonuria (Motulsky et al., 1994).

8.2. Confidentiality

Confidentiality in simpler terms is the state of maintaining the secrecy. It is a condition of nondisclosure or limited disclosure of any sensitive information to others, within confined limits (O'Brien and Chantler, 2003). It can be understood with the help of an example, where, a person undergoing tests may decide the extent to which the physician can share the patient's personal details. The physician may be allowed to disclose the details to the patient's immediate family members but may not be given permission to share them to an insurance company without the patient's consent. Infringement of the confidentiality rights may result either through the deliberate breaches or due to the careless handling of confidential information (Beltran-Aroca et al., 2016).

8.3. Privacy

Once a person undergoes genetic testing, the person has the right to decide the limit to which the genomic data can be accessed. Any type of intrusion without the donor's permission is highly unacceptable and is a crime in legal sphere (Witt and Witt, 2016). The application of DNA testing for studying genetic variation is not just limited to medical field but is also extensively implemented in non-medical fields such as forensics, say, for criminal identification. A set of genetic information that may help detect the genetic predisposition to a disease or may help to distinguish the asymptomatic and symptomatic individuals for medical purposes, may also serve as the deciding factor to rule out the suspects during the convictions in courtrooms. Despite of the vigilance by the Federal Bureau of Investigation's DNA Advisory Board and the other scientific and technical organization over DNA testing within legal setting, there are chances that the private or legal implications of these applications seldom goes overlooked. Some organizations believe that the genetic testing of suspects is an invasion of privacy and personal autonomy. To protect these basic rights of the citizens, the Fourth Amendment of the U.S. Constitution grants its citizens protection against illegal search and seizures (Murphy, 2013). However, the sample retrieved from the crime scene is considered abandoned and is not constitutionally protected (Van Laan, 2017). There is no question of the legal rights of an individual getting violated since the sample collected no longer counts as the property of the individual and upon abandonment; the sample is legally transited from private to legal property (Toom, 2012).

Conclusion and Future Remarks

Since its discovery, DNA fingerprinting persists as a precise and key technique in forensic science to link the suspects with the biological evidence in crimes. Newly renovated techniques have surpassed the traditional methods including RFLP, RAPD, STR and now discerning an individual is possible even through degraded, damaged or festered samples by using mtDNA analysis. In the past decade, methods for collecting and preserving samples have been improved that have raised the probability to perform DNA profiling successfully. Because of the advancement in the past three decades, DNA evidence now stands amongst the most reliable proofs in the courtroom. This not only helps in criminal cases but also for predicting inherited disorders, studying population genetics, figuring out the best match for organ donation, parental tests, identifying victims in disasters and wars and forming lineages. In future, the emergence of nano-technological approaches along with microfluidics platform and universal DNA database would make investigations more simple, rapid and efficient for further processing.

ACKNOWLEDGMENTS

This work was supported by The Puri Foundation for Education in India.

REFERENCES

Aal-Hamdan, H.M., Refaat, A.M., Babu, S.R.,&Choudhry, A.R. (2015).Fetal gender determination through Y-STR analysis of maternal plasma during the third trimester of pregnancy. *Egyptian Journal of Medical Human Genetics*, *16*(1), 47-53.

Abdul-Muneer, P.M. (2014). Application of microsatellite markers in conservation genetics and fisheries management: recent advances in population structure analysis and conservation strategies. *Genetics Research International*, 2014, 691759.

Abu-Salah, K.M., Zourob, M.M., Mouffouk, F., Alrokayan, S.A., Alaamery, M.A., & Ansari, A.A. (2015).DNA-based nanobiosensors as an emerging platform for detection of disease. *Sensors*, *15*(6), 14539-14568.

Al-Jeboory, M.R., Al-Jailawi, M.H., Al-Obaedi, A.M., & Al-Jeboory, S.R. (2015). Improvement of DNA extraction methods by ZnO and TiO_2 nanoparticles. In: *International conference on medical genetics, cellular & molecular biology, pharmaceutical & food sciences (GCMBPF-2015).*

Ambers, A., Votrubova, J., Vanek, D., Sajantila, A., & Budowle, B. (2018). Improved Y-STR typing for disaster victim identification, missing persons investigations, and historical human skeletal remains. *International Journal of Legal Medicine*, doi: 10.1007/s00414-018-1794-8.

Amos, B., Barrett, J., & Dover, G.A. (1991). Breeding behaviour of pilot whales revealed by DNA fingerprinting. *Heredity*, *67*(1), 49-55.

Beltran-Aroca, C.M., Girela-Lopez, E., Collazo-Chao, E., Montero-Pérez-Barquero, M., & Muñoz-Villanueva, M.C. (2016). Confidentiality breaches in clinical practice: what happens in hospitals?. *BMC Medical Ethics*, *17*(1), 52.

Benítez, J. J., Topolancik, J., Tian, H. C., Wallin, C. B., Latulippe, D. R., Szeto, K., Murphy, P.J., Cipriany, B.R., Levy, S.L., Soloway, P.D.& Craighead, H. G. (2012). Microfluidic extraction, stretching and analysis of human chromosomal DNA from single cells. *Lab on a Chip*, *12*(22), 4848-4854.

Bhukya, R. (2016). Extended perfect and imperfect repeats finder using molecular data sequences. *Journal of Theoretical & Applied Information Technology*, *91*(2).

Bieber, F.R., Buckleton, J.S., Budowle, B., Butler, J.M., & Coble, M.D. (2016). Evaluation of forensic DNA mixture evidence: protocol for evaluation, interpretation, and statistical calculations using the combined probability of inclusion. *BMC Genetics*, *17*(1), 125.

Botstein, D., White, R.L., Skolnick, M., & Davis, R.W. (1980). Construction of a genetic linkage map in man using restriction fragment length polymorphisms. *American Journal of Human Genetics*, *32*(3), 314.

Brettschneider, R.(1998). RFLP analysis. In: *Molecular Tools for Screening Biodiversity*, 85-95. Springer, Dordrecht.

Bruijns, B., van Asten, A., Tiggelaar, R., & Gardeniers, H. (2016). Microfluidic devices for forensic DNA analysis: A review. *Biosensors*, *6*(3), 41.

Budowle, B., Allard, M.W., Wilson, M.R., & Chakraborty, R. (2003). Forensics and mitochondrial DNA: applications, debates, and foundations. *Annual Review of Genomics & Human Genetics*, *4*(1), 119-141.

Budowle, B., Wilson, M.R., DiZinno, J.A., Stauffer, C., Fasano, M.A., Holland, M.M., & Monson, K.L. (1999). Mitochondrial DNA regions HVI and HVII population data. *Forensic Science International*, *103*(1), 23-35.

Buscot, F., Wipf, D., Di Battista, C., Munch, J.C., Botton, B., & Martin, F. (1996). DNA polymorphism in morels: PCR/RFLP analysis of the ribosomal DNA spacers and microsatellite-primed PCR. *Mycological Research*, *100*(1), 63-71.

Butler, J.M., Coble, M.D., & Vallone, P.M. (2007). STRs vs. SNPs: thoughts on the future of forensic DNA testing. *Forensic Science, Medicine, and Pathology*, *3*(3), 200-205.

Cai, M., Li, F., Zhang, Y., & Wang, Q. (2010). One-pot polymerase chain reaction with gold nanoparticles for rapid and ultrasensitive DNA detection. *Nano Research*, *3*(8), 557-563.

Carroll, D. (2016). Genome editing: Progress and challenges for medical applications. *Genome Medicine*, *8*(1), 120.

Carroll, D. (2017). Genome editing: Past, present, and future. *The Yale Journal of Biology and Medicine*, *90*(4), 653.

Chawla, P. (1998). *The Detective DNA*. NISCAIR, CSIR, New Delhi.

Cherstvy, A.G. (2013). Detection of DNA hybridization by field-effect DNA-based biosensors: mechanisms of signal generation and open questions. *Biosensors and Bioelectronics*, *46*, 162-170.

Collins, F.S., Guyer, M.S., & Chakravarti, A. (1997). Variations on a theme: cataloging human DNA sequence variation. *Science*, *278*(5343), 1580-1581.

Du, X., An, H., Jin, B., Meng, L., & Liu, Q. (2014). Carbon nanotubes altering specificity of repeated PCR and DNA integrity properties. *Journal of Nanoscience and Nanotechnology*, *14*(7), 5547-5551.

Dubey, R.C. (2014). *Advanced Biotechnology*. S. Chand Publishing.

Ellegren, H. (2000). Heterogeneous mutation processes in human microsatellite DNA sequences. *Nature Genetics*, *24*(4), 400.

Fan, H., & Chu, J.Y. (2007). A brief review of short tandem repeat mutation. *Genomics, Proteomics & Bioinformatics*, *5*(1), 7-14.

Gefrides, L.A., & Welch, K.E. (2006). Serology and DNA. In *The Forensic Laboratory Handbook*, 1-33. Humana Press.

Gill, P., Whitaker, J., Flaxman, C., Brown, N., & Buckleton, J. (2000). An investigation of the rigor of interpretation rules for STRs derived from less than 100 pg of DNA. *Forensic Science International*, *112*(1), 17-40.

Gupta, S., Ramesh, K., Ahmed, S., & Kakkar, V. (2016). Lab-on-Chip Technology: A Review on Design Trends and Future Scope in Biomedical Applications. *International Journal of Bio-science and Bio-technology*, *8*(5), 311-322.

Gymrek, M., Willems, T., Guilmatre, A., Zeng, H., Markus, B., Georgiev, S., Daly, M.J., Price, A.L., Pritchard, J.K., Sharp, A.J., & Erlich, Y. (2016). Abundant contribution of short tandem repeats to gene expression variation in humans. *Nature Genetics*, *48*(1), 22.

Hajihosseini, S., Nasirizadeh, N., Hejazi, M.S., & Yaghmaei, P. (2016). A sensitive DNA biosensor fabricated from gold nanoparticles and graphene oxide on a glassy carbon electrode. *Materials Science and Engineering: C*, *61*, 506-515.

Ham, S.K., Kim, S.Y., Seo, B.Y., Woo, K.M., Lee, S.H., & Choi, C.Y. (2016). Differential pre-amplification of STR loci for fragmented forensic DNA profiling. *Electrophoresis*, *37*(22), 3002-3009.

Holt, C.L., Buoncristiani, M., Wallin, J.M., Nguyen, T., Lazaruk, K.D., & Walsh, P.S. (2002). TWGDAM validation of AmpFlSTR PCR amplification kits for forensic DNA casework. *Journal of Forensic Science*, *47*(1), 66-96.

Lander, E.S., Linton, L.M., Birren, B., Nusbaum, C., Zody, M.C., Baldwin, J., et al. (2001). Initial sequencing and analysis of the human genome. *Nature*, *409*(6822), 860.

Jeffreys, A.J., Brookfield, J.F., & Semeonoff, R. (1985). Positive identification of an immigration test-case using human DNA fingerprints. *Nature*, *317*(6040), 818.

Jeffreys, A.J., & Pena, S.D.J. (1993). Brief introduction to human DNA fingerprinting. In *DNA Fingerprinting: State of the Science*, 1-20. Birkhäuser, Basel.

Jobling, M.A. (2013). Curiosity in the genes: the DNA fingerprinting story. *Investigative Genetics*, *4*, 20.

Kambli, P., & Kelkar-Mane, V. (2016). Nanosized Fe3O4 an efficient PCR yield enhancer - Comparative study with Au, Ag nanoparticles. *Colloids and Surfaces B: Biointerfaces*, *141*, 546-552.

Kayser, M. (2015). Forensic DNA phenotyping: predicting human appearance from crime scene material for investigative purposes. *Forensic Science International: Genetics*, 18, 33-48.

Kayser, M. (2017). Forensic use of Y-chromosome DNA: a general overview. *Human Genetics*, *136*(5), 621-635.

Kayser, M., & Schneider, P.M. (2009). DNA-based prediction of human externally visible characteristics in forensics: motivations, scientific challenges, and ethical considerations. *Forensic Science International: Genetics*, *3*(3), 154-161.

Khairkar, S.R., Palaskar, S.G., Kulkarni, K.V., & Malve, M.K. (2015).Importance of ABO Blood Grouping in the Era of DNA Analysis-A Case solved by ABO Blood Grouping Method. *Journal of Forensic Research*, *6*(4), 1.

Kim, K.S., & Sappington, T.W. (2013). Microsatellite data analysis for population genetics. In *Microsatellites,* 271-295. Humana Press, Totowa, NJ.

Kline, M.C., Vallone, P.M., Redman, J.W., Duewer, D.L., Calloway, C.D., & Butler, J.M. (2005). Mitochondrial DNA typing screens with control region and coding region SNPs. *Journal of Forensic Science*, *50*(2), 377-385.

Kobilinsky, L.F., Liotti, T.F., & Oeser-Sweat, J. (2005). *DNA: Forensic and Legal Applications*, *155*. Wiley-Interscience.

Koonin, E.V., &Novozhilov, A.S. (2009). Origin and evolution of the genetic code: the universal enigma. *IUBMB Life*, *61*(2), 99-111.

Koops, B.J., Prinsen, M., & Schellekens, M. (2006). Wanted: A Tall Blond Dutchman-Does the Netherlands Set the Stage in Regulating Forensic DNA Phenotyping. *Tilburg Foreign Law Review*, *13*, 206.

Kumari, N., & Thakur, S.K. (2014).Randomly amplified polymorphic DNA-A brief review.*American Journal of Animal and Veterinary Sciences*, *9*(1), 6-13.

Ladd, C., Lee, H.C., Yang, N., & Bieber, F.R., 2001. Interpretation of complex forensic DNA mixtures.*Croatian Medical Journal*, *42*(3), 244-246.

Lander, E.S. (2011). Initial impact of the sequencing of the human genome. *Nature*, *470*(7333), 187.

Li, M., Lin, Y.C., Wu, C.C., & Liu, H.S. (2005). Enhancing the efficiency of a PCR using gold nanoparticles. *Nucleic Acids Research*, *33*(21), e184.

Li, Z., Chen, F., Huang, C., Zheng, W., Yu, C., Cheng, H., & Zhou, R. (2017). Genome-wide mapping and characterization of microsatellites in the swamp eel genome. *Scientific Reports*, *7*(1), 3157.

Lodha, A., Ansari, N., Shah, S., Rao, M.V., & Menon, S.K. (2017). Isolation of PCR ready-human DNA using copper nanoparticles from skeletal remains. *Forensic Science International*, *270*, 146-152.

Lorenz, T.C. (2012). Polymerase chain reaction: basic protocol plus troubleshooting and optimization strategies. *Journal of Visualized Experiments*, *63*, 3998.

Lynch, M., & Milligan, B.G. (1994). Analysis of population genetic structure with RAPD markers. *Molecular Ecology*, *3*(2), 91-99.

MacLean, C.E. (2013). Creating a wanted poster from a drop of blood: using DNA phenotyping to generate an artist's rendering of an offender based only on DNA shed at the crime scene. *Hamline Law Review*, *36*, 357.

Mahmoud, M.A., Afifi, M.I., Al-Othman, M.R., & EL-AZIZ, A.R.M.A.B.D. (2017). Addition of silver and titanium dioxide nanoparticles to real-time PCR for the detection *Aspergillus Spp.* and aflatoxin genes. *Digest Journal of Nanoparticles and Biostructures*, *12*(4), 1041-1048.

McCord, B. (2006). Nanotechnology and Its Potential in Forensic DNA Analysis. *Promega*, 7-9.

McEwen, J.E. (1995). Forensic DNA data banking by state crime laboratories. *American Journal of Human Genetics*, *56*(6), 1487.

Moretti, T.R., Baumstark, A.L., Defenbaugh, D.A., Keys, K.M., Smerick, J.B., & Budowle, B. (2001). Validation of short tandem repeats (STRs) for forensic usage: performance testing of fluorescent multiplex STR systems and analysis of authentic and simulated forensic samples. *Journal of Forensic Science*, *46*(3), 647-660.

Motulsky, A.G., Holtzman, N.A., Fullarton, J.E., & Andrews, L.B. eds. (1994). Assessing genetic risks: implications for health and social policy (Vol. 1). *National Academies Press.*

Mukherjee, B.(2013). A scientometric profile of Prof. Lalji Singh as seen through Web of Science and Scopus. *Annals of Library and Information Studies,60*(3), 195-203.

Murphy, E. (2013). Legal and ethical issues in forensic DNA phenotyping. *New York University Public Law and Legal Theory Working Papers.* Paper 415.

Murray-McIntosh, R.P., Scrimshaw, B.J., Hatfield, P.J., & Penny, D. (1998). Testing migration patterns and estimating founding population size in Polynesia by using human mtDNA sequences. *Proceedings of the National Academy of Sciences*, *95*(15), 9047-9052.

Nims, R.W., Sykes, G., Cottrill, K., Ikonomi, P., & Elmore, E. (2010). Short tandem repeat profiling: part of an overall strategy for reducing the frequency of cell misidentification. *In Vitro Cellular & Developmental Biology-Animal*, *46*(10), 811-819.

O'brien, J., & Chantler, C. (2003). Confidentiality and the duties of care. *Journal of Medical Ethics*, *29*(1), 36-40.

Patel, N., Gautaman, V.K., & Jangir, S. (2013). The role of DNA in criminal investigation–Admissibility in Indian legal system and future perspectives. *International Journal of Humanities and Social Science Invention*, *2*, 15-21.

Paul, T., Basu, S., & Sarkar, K. (2014). SPION-mediated soil DNA extraction and comparative analysis with conventional and commercial kit-based protocol. *3 Biotech*, *4*(6), 669-677.

Paun, O., & Schönswetter, P. (2012). Amplified fragment length polymorphism: an invaluable fingerprinting technique for genomic, transcriptomic, and epigenetic studies. In *Plant DNA Fingerprinting and Barcoding*, 75-87. Humana Press.

Peterson, J., Sommers, I., Baskin, D., & Johnson, D. (2010). The role and impact of forensic evidence in the criminal justice process. *National Institute of Justice*, 1-151.

Prinz, M. (2003). Advantages and Disadvantages of Y-Short Tandem Repeat Testing in Forensic Casework. *Forensic Science Review*, *15*(2), 191-198.

Ramnarayan, B.K., Manjunath, M., & Joshi, A.A. (2013). ABO blood grouping from hard and soft tissues of teeth by modified absorption-elution technique. *Journal of Forensic Dental Sciences*, *5*(1), 28-34

Rasmussen, H.B. (2012). Restriction fragment length polymorphism analysis of PCR-amplified fragments (PCR-RFLP) and gel electrophoresis-valuable tool for genotyping and genetic fingerprinting. In *Gel electrophoresis-principles and basics*. InTech.

Rehman, A., Sarwar, Y., Raza, Z.A., Hussain, S.Z., Mustafa, T., Khan, W.S., Ghauri, M.A., Haque, A., & Hussain, I. (2015). Metal nanoparticle assisted polymerase chain reaction for strain typing of *Salmonella Typhi*. *Analyst*, *140*(21), 7366-7372.

Resnik, D.B., & Vorhaus, D.B. (2006). Genetic modification and genetic determinism. *Philosophy, Ethics, and Humanities in Medicine*, *1*(1), 9.

Roberts, R.J., & Murray, K. (1976). Restriction endonuclease. *CRC Critical Reviews in Biochemistry*, *4*(2), 123-164.

Roewer, L. (2013). DNA fingerprinting in forensics: past, present, future. *Investigative Genetics*, *4*(1), 22.

Rolf, B., Keil, W., Brinkmann, B., Roewer, L., & Fimmers, R. (2001). Paternity testing using Y-STR haplotypes: assigning a probability for paternity in cases of mutations. *International Journal of Legal Medicine*, *115*(1), 12-15.

Růžička, M., Kulhánek, P., Radová, L., Čechová, A., Špačková, N., Fajkusová, L., & Réblová, K. (2017). DNA mutation motifs in the genes associated with inherited diseases. *PloS One*, *12*(8), e0182377.

Saad, R. (2005). Discovery, development, and current applications of DNA identity testing. In *Baylor University Medical Center Proceedings*, *18*(2), 130-133.

Saeed, A.A., Sánchez, J.L.A., O'Sullivan, C.K., & Abbas, M.N. (2017). DNA biosensors based on gold nanoparticles-modified graphene oxide for the detection of breast cancer biomarkers for early diagnosis. *Bioelectrochemistry*, *118*, 91-99.

Saiyed, Z.M., Ramchand, C.N., & Telang, S.D. (2008). Isolation of genomic DNA using magnetic nanoparticles as a solid-phase support. *Journal of Physics: Condensed Matter*, *20*(20), 204153.

Saiyed, Z.M., & Ramchand, C.N. (2007). Extraction of genomic DNA using magnetic nanoparticles (Fe3O4) as a solid-phase support. *American Journal of Infectious Diseases, 3*(4), 225-229.

Saks, M.J. and Koehler, J.J. (2005).The coming paradigm shift in forensic identification science. *Science*, *309*(5736), 892-895.

Sensabaugh, G.F. (1982). Isozymes in forensic science. *Isozymes Current Topics in Biological and Medical Research*, *6*, 247-282.

Serpionov, G.V., Alexandrov, A.I., Antonenko, Y.N., & Ter-Avanesyan, M.D. (2015). A protein polymerization cascade mediates toxicity of non-pathological human huntingtin in yeast. *Scientific Reports*, *5*, 18407.

Shabbir, M.A., Hao, H., Shabbir, M.Z., Wu, Q., Sattar, A., & Yuan, Z. (2016). Bacteria vs. bacteriophages: parallel evolution of immune arsenals. *Frontiers in Microbiology*, *7*, 1292.

Shapiro, E.D., Reifler, S., & Psome, C.L. (1992). The DNA Paternity Test: Legislating the Future Paternity Action. *JL & Health*, *7*, 1-47.

Shrivastava, P., Trivedi, V.B., Singh, A.K., & Mishra, N. (2012). Application of DNA fingerprinting technology in forensic investigation. *International Journal of Scientific and Research Publications, 2*(10), 1-4.

Siddiquee, S., Rovina, K., Yusof, N.A., Rodrigues, K.F., & Suryani, S., 2014. Nanoparticle-enhanced electrochemical biosensor with DNA immobilization and hybridization of Trichodermaharzianum gene. *Sensing and Bio-sensing Research*, *2*, 16-22.

Singh, V., Braddick, D., & Dhar, P.K. (2017). Exploring the potential of genome editing CRISPR-Cas9 technology. *Gene*, *599*, 1-18.

Singh, V., Gohil, N., Ramírez García, R., Braddick, D., & Fofié, C.K. (2018). Recent advances in CRISPR-Cas9 genome editing technology for biological and biomedical investigations. *Journal of Cellular Biochemistry*, *119*(1), 81-94.

Sowmyya, T. (2016). Touch DNA: An Investigative Tool in Forensic Science. *International Journal of Current Research*, *8*(2), 26093-26097.

Strehle, M.M., Purfeerst, E., & Christensen, A.C. (2018). A rapid and efficient method for enriching mitochondrial DNA from plants. *Mitochondrial DNA Part B*, *3*(1), 239-242.

Sullivan, K.M. (1994). Forensic applications of DNA fingerprinting. *Molecular Biotechnology*, *1*(1), 13-27.

Tang, H., & Nzabarushimana, E. (2017). STRScan: targeted profiling of short tandem repeats in whole-genome sequencing data. *BMC Bioinformatics*, *18*(11), 398.

Thangaraj, K., Chaubey, G., Kivisild, T., Reddy, A.G., Singh, V.K., Rasalkar, A.A., & Singh, L. (2005). Reconstructing the origin of Andaman Islanders. *Science*, *308* (5724), 996-996.

Toom, V. (2012). Bodies of science and law: Forensic DNA profiling, biological bodies, and biopower. *Journal of Law and Society*, *39*(1), 150-166.

Urquhart, A., Kimpton, C.P., Downes, T.J., & Gill, P. (1994). Variation in short tandem repeat sequences—a survey of twelve microsatellite loci for use as forensic identification markers. *International Journal of Legal Medicine*, *107*(1), 13-20.

Van Laan, M.C. (2017). The genetic witness: forensic DNA phenotyping. *Journal of Emerging Forensic Sciences Research*, *2*(2), 33-52.

Vanzha, E., Pylaev, T., Khanadeev, V., Konnova, S., Fedorova, V., Khlebtsov, N. (2016). Gold nanoparticle-assisted polymerase chain reaction: effects of surface ligands, nanoparticle shape and material. *RSC Advances*, *6*(111), 110146-110154.

Vidaki, A., & Daniel, B. (2013). Forensic DNA methylation profiling—potential opportunities and challenges. *Forensic Science International: Genetics*, *7*(5), 499-507.

von Wurmb-Schwark, N., Petermann, S., & Wegener, R. (2003). Y-STR typing in forensic analysis. *International Congress Series*, *1239*, 487-490.

Vos, P., Hogers, R., Bleeker, M., Reijans, M., Lee, T.V.D., Hornes, M., Friters, A., Pot, J., Paleman, J., Kuiper, M., & Zabeau, M. (1995). AFLP: a new technique for DNA fingerprinting. *Nucleic Acids Research*, *23*(21), 4407-4414.

Vu, B.V., Litvinov, D., & Willson, R.C. (2008). Gold nanoparticle effects in polymerase chain reaction: favoring of smaller products by polymerase adsorption. *Analytical Ahemistry*, *80*(14), 5462-5467.

Watson, J.D. (2012). *The polymerase chain reaction*. Springer science & business media.

Wickenheiser, R.A. (2002). Trace DNA: a review, discussion of theory, and application of the transfer of trace quantities of DNA through skin contact. *Journal of Forensic Science*, *47*(3), 442-450.

Williamson, A.L. (2012). Touch DNA: forensic collection and application to investigations. *Journal of the Association for Crime Scene Reconstruction*, *18*(1), 1-5.

Witt, M.M., & Witt, M.P. (2016). Privacy and confidentiality measures in genetic testing and counselling: arguing on genetic exceptionalismagain?. *Journal of Applied Genetics*, *57*(4), 483-485.

Wynbrandt, F., & Chisum, W.J. (1971). Determination of the ABO blood group in hair. *Journal of the Forensic Science Society*, *11*(3), 201-204.

Yeow, J.T., & Wan, W. (2010). The effects of nanoparticles on polymerase chain reaction. In *Nano/Micro Engineered and Molecular Systems (NEMS), 2010 5th IEEE international conference,* 855-859.

Zhang, Z., Shen, C., Wang, M., Han, H., & Cao, X. (2008). Aqueous suspension of carbon nanotubes enhances the specificity of long PCR. *BioTechniques*, *44*(4), 537-545.

Zhou, H., Liu, J., Xu, J.J., & Chen, H.Y. (2011). Ultrasensitive DNA detection based on Au nanoparticles and isothermal circular double-assisted electrochemiluminescence signal amplification. *Chemical Communications*, *47*(29), 8358-8360.

Zhou, Z., Kadam, U.S., & Irudayaraj, J. (2013). One-stop genomic DNA extraction by salicylic acid-coated magnetic nanoparticles. *Analytical Biochemistry*, *442*(2), 249-252.

In: Introduction of Forensic Nanotechnology … ISBN: 978-1-53616-040-6
Editors: R. K. Shukla and A. Pandya © 2019 Nova Science Publishers, Inc.

Chapter 7

IDENTIFICATION OF FOOD ADULTERANTS USING NANOTECHNOLOGY

Ramesh Chaudhari and Ashutosh Kumar*
Biological and Life Sciences, School of Arts and Sciences,
Ahmedabad University, Ahmedabad, Gujarat, India

ABSTRACT

Food safety plays an important role in community health and thus to society as a whole. Foodborne disease associated with pathogens, toxins or other food contaminants poses a severe health risk across the globe. Food adulteration makes food unsafe due to the presence of adulterants such as melamine, foodborne pathogenic bacteria, toxins such as cholera, aflatoxin amongst others. The present chapter is focused on the potential role of a nanomaterial, which is currently being used in various biosensors, for the detection of various chemical contaminants, toxins, and pathogens and discusses the advantages of nanotechnology-based detection system over a traditional system.

Keywords: food adulteration, food safety, health risk, nanomaterial, nanosensors

* Corresponding Author's E-mail: ashutosh.kumar@ahduni.edu.in.

1. Introduction

Food is the vital requisite of life and the most important thing for a living being to survive after air and water. The food contains various micronutrients (e.g., vitamins and minerals) which are required in small quantity, as well as macronutrients (e.g., carbohydrates, protein, and fat) which are required in larger quantity [1]. Food is essential for our body to perform various functions for example,

1. **For energy:** Body operates using energy supplied from the food in the form of glucose which is utilized in mitochondria to produce adenosine triphosphate (ATP). It is also known as the "molecular unit of currency", as it provides energy to drive many processes in living cells.
2. **As a medicine:** There are kinds of a diseases which can occur due to lack of food or nutrition's such as "rickets" due to the deficiency of the vitamin D and calcium; "scurvy" deficiency of the vitamin C or ascorbic acid; "Beri Beri" lack of the vitamin B1 or thiamine among others. Malnutrition and deficiency of these vital elements can lead to a serious diseases and conditions. Additionally, the excess of a food can also lead to serious health conditions like excess weight gain, overweight and obesity, physical inactivity and increased risk of heart & kidney disease. Food also boosts up the immunity and protects from various common infections. Phytochemicals, plant-derived, non-nutritive and biologically active chemicals function in the body to prevent the onset of certain non-communicable diseases.
3. **For a drug to act:** Drug act well when taken with sufficient food. Food gives stamina to the body to accommodate a drug safely and also helps to metabolize it after its action. Some diseases required sufficient food supply along with a drug to overcome it like sunstroke, tuberculosis, gastric ulcers, muscle pains, skin diseases, etc. Oil soluble drugs can easily absorb in the presence of a fatty food. Food also acts as an enzyme inducer, as a result, various

enzymes are generated which helps in metabolizing toxic drugs in the liver.

Apart from this, food also play a role in maintaining body shape, stress reliever, etc. [2].

1.1. Importance of Safe Food

Contaminated and adulterated foods are key sources for some food-borne diseases in the nation. Various microorganisms and natural enzymes are primarily responsible for the spoilage of food. Moreover, adulterants, insects, natural toxins, rodents and various chemical residues beyond approved levels, make the food objectionable. In addition environmental conditions like a temperature, moisture, storage time also affect the quality of the food.

Foods may be mixed with non-food material or cheap quality product. Inferior, musty or poor quality food is made eye-catching and renewed by adding harmful colors or other chemicals. Commonly adulterated food substances are milk and milk products, cereals, pulses, and their products, edible oils and spices. The different classes of adulterants contain non-permitted colors like metanil yellow; non-edible oils like castor oil; cheaper agricultural food like various thickeners in a milk powder; extraneous material like husk, sand, and sawdust; and metal impurities like aluminum or iron filings. Intake of adulterated foods could lead to disease outbreaks of epidemic proportions. Procurement from a reliable and reputed source, careful examination of foods before buying and insisting on certified products can reduce the risk of food adulteration.

1.2. Food Adulteration and Toxicity

Food adulteration has a long history. Food adulteration is not a novel and rare concern in human society; indeed, it can be traced back to the 18th

and 19th centuries and was common in the European and American regions [3]; however, development in agriculture and globalization have helped in spreading the problem across the globe [4]. Food Adulteration refers to the process by which the quality or the nature of a given food is reduced through the addition of an adulterants or removal of a vital substance. According to Food Safety and Standards Act of India (FSSAI), "adulterant" means any material which is or could be employed for making the food unsafe or sub-standard or misbranded or containing extraneous matter [3].

Animals, humans, plants, and microorganisms are continuously being exposed to various toxic chemicals that are released from many different sources and may enter the living organisms by different routes. It is well known that excessive usage of a toxic chemical such as heavy metal ions, food additives, and pesticides bring severe health problems to living organisms and pose long term threat to environmental sustainability. The list of toxic chemicals is very long. Food and water that we consume on a daily basis are even a contained unsafe concentration of these chemicals. Generally, these toxic chemicals have entered the environment through natural and anthropogenic sources. They get into the human food chain from the environment. Once they enter into the biological system they disturb the biological processes, leading in some cases to fetal deaths [4]. The major causes of food adulteration are toxic chemicals which are responsible for various foodborne disease outbreaks [5]. Toxic chemicals may be classified according to their function and effects, such as mutagens, carcinogens, food additives, heavy metals, metal carbonyls, pesticides, and explosive compounds. Studies have demonstrated that a food contamination due to chemicals, adulterated from a natural or an anthropogenic source is a significant source of foodborne illness and it poses a severe risk to human health. Also, this has been reported that the chemical contamination of food can affect health not only after a single massive exposure but also after continuous exposure to low doses of a toxic chemical that can be related to several chronic diseases, including some type of a cancer and serious hormonal dysfunctions.

There are several reasons for the adulteration of food [6]. Food processing involves various stages, and in each stage of processing, incursion of a chemical contaminant in food is very likely. Shipping of food under poor hygiene condition can also lay the foundation for food contamination [7]. Also, to improve the shelf-life of food, some chemicals are added purposefully during the preparation of a food.

1.3. Type of Food Adulteration

Adulteration includes the addition of an unnecessary, unsafe, inefficient substance to food which affects the quality of a food. The food can be adulterated due to the addition of contaminants during food processing, the addition of unapproved adulterants as a food additives and migrant from packaging material [8]. Food processing contaminants are the various compounds formed during the food processing like baking, roasting, canning, heating, fermentation, or hydrolysis which are undesirable [9]. Some harmful chemicals can also leach into the food from the packaging material because of their direct contact with food.

It can be classified in various categories as follow:

1.3.1. Unintentional Adulteration

During the processing of food some substances are added because of the ignorance, carelessness or may be due to the lack of hygiene or improper facility. This can lead to a contamination of food by bacteria or fungi, dust and stone or by the packing material. These kinds of adulterations can be considered as an acquired type of adulteration. Some chemicals, organic compounds or radicals can be present in the food naturally, which are known as an inherent adulteration, for example, toxic varieties of mushroom, pulse, seafood, and other vegetables.

1.3.2. Intentional Adulteration

Intentional adulteration of food includes adulteration done by adding knowingly to get a better reward. Melamine contains high nitrogen level; it is illegally added to milk products to generate a false high reading in the quantification of protein content based on total nitrogen. On the other hand, bisphenol A is one of the food contaminants, extensively used as a material for the production of epoxy resins and polycarbonate plastics. It has been reported to have detrimental reproductive effects on wildlife and humans by affecting their endocrine function and may disrupt growth and development by disturbing the different cellular and developmental processes.

1.3.3. Metallic Contamination

Metallic contamination includes the addition of different types of metals or metal compound in food. This contamination could be intentional or unintentional. Lead, cadmium, arsenic, and mercury are the most toxic heavy metals and can lead to multiple organ failure and many other complicated issues to the living organisms.

1.3.4. Microbial Contamination

Microbial contamination includes the spoilage of food due to the microbes entering food from various sources. These sources could be, the time of harvesting, packaging, storage, transportation, processing, etc.

Food packaging methods make the use of a various additives such as antioxidants, slipping agents, stabilizers, and plasticizers to enhance the packaging material properties. Nevertheless, direct or indirect contact of food with packaging material can lead to the leaching of their harmful substances in the food [10]. To overcome this, the internal surfaces of cans are usually covered with varnishes like epoxy resins, but even the slight by-product from these resins like cyclo-di- BADGE, bisphenol A, or bisphenol A diglycidyl ether (BADGE) can be leached to food. Such compounds are identified as endocrine disruptors [(11)] and should not be used.

1.4. Health Risks of Adulterated Food

Food adulteration can lead to various complications ranging from minor to severe life-threatening conditions like food poisoning, vision complication, diarrhea, and skin related problems. Asthma, skin diseases and cancer triggered due to the consumption of meat, fish, milk or fruit that adulterated using chemicals like formalin are the usually and frequently seen examples.

Sometimes food adulteration shows immediate adverse effects due to the high sensitivity of human health like diarrhea, vomiting, and dysentery. For instance, if the coffee powder is mixed with a date seed powder or a tamarind can cause diarrhea [12]. Adulteration in a dairy and bakery products causes abdominal cramps and vomiting. In the case of a milk and meat if not processed properly it may cause food poisoning and other infections along with a fever and chills. Apart from these immediate effects, there may be long-term adverse effects like peptic ulcers, heart disease, kidney and liver damage can also be occurred. These can be observed due to excess amount of permitted preservatives like coloring agents, urea, burnt engine oil, and calcium bromide, etc.

In the United States, annual foodborne diseases number about 48 million illnesses [13]. In 2013, there were more than 11,000 foodborne infections were reported according to the US Centre for Disease Control and Prevention [14]. The signs of the foodborne sickness due to chemical adulterants range from minor gastroenteritis to severe cases of a renal, hepatic, and neurological disorders. In the United States, around 1500 outbreaks of foodborne diseases were reported between 2009 and 2010 which lead to around twenty-nine thousand illness cases and twenty-three deaths [15]. The recent development of the industries and the subsequent environmental pollution made food adulteration more thoughtful [16]. Besides that, heavy metal and pesticides contaminated food uptake could cause gastrointestinal infection [16]. For example, in Nigeria, around 400 to 500 children were died due to acute lead poisoning after ingestion of food contaminated with lead [17]. A substantial percentage of China's milk supplies were found to have adulterated using melamine in 2008. This resulted in a famous 2008 Chinese milk scandal in which around 8 children were died and estimated around thousands were harmed. Similarly, in 2012, Food Safety Standards Authority of India (FSSAI) conducted a study across 33 states of India and found that milk in India is tainted with detergent, fat and even urea, as well diluted with water [12]. Heavy metal contamination such as Zn, Pb, and Ni was found above the safe limits in vegetables grown in wastewater irrigated areas of Varanasi [18]. In 1987, several thousand people in Kashmir valley were affected by an outbreak of trichothecene Mycotoxicosis associated with consumption of mold-damaged wheat products in Kashmir valley [19].

2. Methods for Detecting Adulterants

Substitution of any constituent in food is either a total substitution or a partial substitution. Detection of a partial substitution is difficult, as the identity of partial adulterants becomes the primary problem instead of the concentration of adulterants. Several methods, based on the morphological/ anatomical description, organoleptic indications (odor, color, texture) and chemical testing have been developed for qualitative assessment of adulterants in a traded food product [20].

Followings are the basic approaches used as of now to indicate the presence of food adulterant: -

- By demonstrating the presence of external ingredient or a marker in the product
- By comparing the level of a constituent from its standard value and
- By indicating a doubtful chemical profile of the food

Among these, the simplest and widely implemented approach for detection of food adulterant is based on the identification of adulterant in the food from external sources [21]. The detection method of the adulterant usually varies according to the type and nature of adulterant to be detected. Major methods used for the adulterants detection include physical, chemical, and analytical, DNA based molecular methods and most recently the metal nanoparticle based sensing approach. (Figure 1) [22].

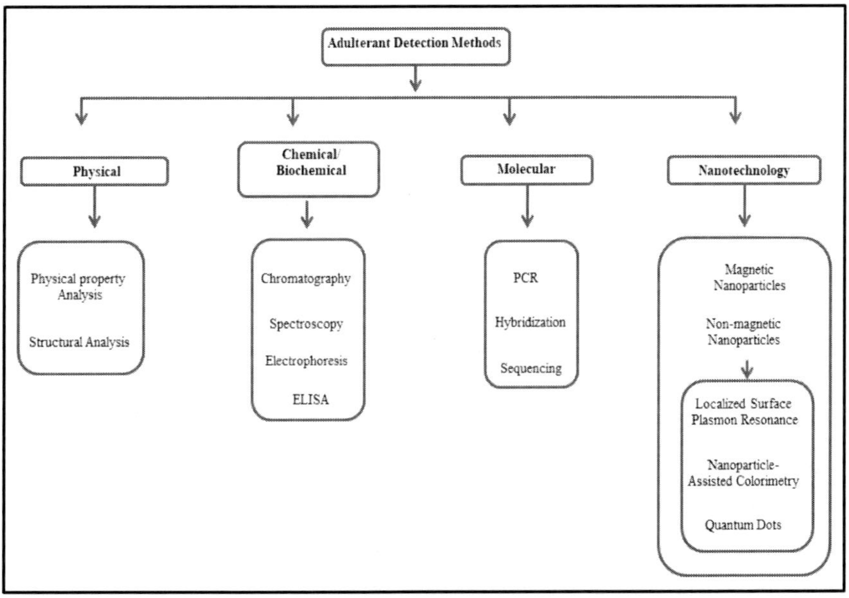

Figure 1. Various methods for the detection of adulterants.

2.1. Physical Methods

Physical methods for detection of adulteration include the examination of physical parameters of the food like, solubility, texture, bulk density, morphology, and micro or macroscopic analysis of food product, etc. The major problem associated with these types of test methods is their qualitative detection. In case of an herb or a medicinal plant identification and authentication, chemical profiling along with macro and microscopy examination is required as per the guideline of pharmacopeia regulatory authorities [23]. Microscopy is very useful in the case of microbial contamination for visual analysis mostly for fungi [24]. According to the FSSA, microscopic inspection of some spices may lead to the detection of extraneous starch like chilies, coriander, cumin and cloves [25].

Additionally, microscopic methods are not up to the mark for detection of adulterants in liquids except for honey. Cane sugar and its products based adulteration in honey can microscopically be detected by the presence of sclerous rings, parenchyma cells, and other cane sugar constitutive cells [26]. With the arrival of advanced microscopes (i.e., electron microscopy), it is now feasible to identify the botanical origin of honey by analyzing a pollen pattern from honey by scanning electron microscope (SEM) [27]. SEM has improved detection capability of microscopy. However, the utilization of SEM for routine examination is not feasible due to the high maintenance and affordability.

2.2. Chemical/Biochemical Methods

Many chemical and biochemical methods based on high-end techniques such as spectroscopy, chromatography, electrophoresis, and immunological reactions have been developed for the detection of an adulterant. These techniques are more sensitive and accurate than the physical methods, but their application at large scale is limited due to its cost and requirement of skilled personnel for its operation. The basic analytical approach involves extraction using appropriate solvent, cleanup for removal of interfering matrix constituent, separation using chromate-graphic techniques and selective detection of specific adulterant using spectroscopy. High-performance liquid chromatography (HPLC) is the most widely used analytical technique. The ability of HPLC to separate various chemical ingredients from mixture makes it a versatile tool to use in quality control, characterizing food products and to detect adulteration. HPLC is used in the detection of various adulterants for example detection of flavones glycosides and polymethoxylated flavones from citrus juices [28], hazelnut oil adulteration in olive oil [29], quince jams with apple or pear puree [30], black tea liquors with phenolic pigments [31], wines adulteration with proline isomers and amino acids [32].

Separation of volatile organic compounds can be done with the help of gas chromatography (GC). GC along with mass spectroscopy (MS) and

Fourier transform infrared spectroscopy (FTIR) has been generally implemented for the detection of an adulterant from the sample in a non-destructive manner. GC is usually used to differentiate varieties of the same product, authentication, and identification of organic compounds and the detection of an adulteration. Gas chromatography has been used to discriminate a wines from the same reason. Various volatile compounds like 2-methyl-1-propanol, 2-propane-1-o1, 3-methyl-1-butanol, and 1-propanol were quantified using GC or GCMS, which provides 30 different physicochemical parameters that help in pattern classification [33]. Near infrared spectroscopy (NIR) is a spectroscopic technique that rapidly detects adulterants in raw material but cannot identify the contaminant. Detection of adulteration or fraud in soy-based products used for animal feed is possible with the NIR spectroscopy [34]. Adulterant detection along with its structural identification can be obtained with the use of nuclear magnetic resonance (NMR) spectroscopy. Differentiation between an adulterated and unadulterated sample can be done using FTIR due to detailed spectral inspection [35]. FTIR spectroscopy has excellent potential to detect an adulterant from milk and it can also be useful in the food industry to takeover slow and low efficient techniques [36]. Handheld Raman device provides a faster measurement compared to FTIR but unable to distinguish between an adulterated and unadulterated product [37]. In oil gamma oryzanol, content (%) has been determined by taking absorbance at 315nm (λ_{max}) with the spectrometric technique [38]. Combination of different techniques like chromatography and spectroscopy has a high potential for detection of adulterant e.g., GC-MS for honey adulteration [39], Solid phase micro-extraction (SPME)-GC-MS for ground roasted coffee adulteration in rostered barley [40], LC-MS-MS for an adulteration detection in herbal remedies [41]. Recently FTIR spectroscopy in combination with chemometric methods showed the potential of rapid detection of adulterants [36]. According to World Health Organization (WHO), Inductive Coupled Plasma - Mass Spectrometry (ICP-MS) can be used as the confirmative technique for an investigation of the lead in all food [42].

Electrophoresis is also used in the detection like in the case of milk and dairy products presence of additional whey can be detected and also can be quantified [43]. Adulteration of cow milk in a goat milk product and adulteration in basmati rice is detected using the capillary zone electrophoresis [44, 45].

Amongst the immunological methods, enzyme-linked immune sorbent assay (ELISA) is a commonly used a form of the immunoassays in adulterant detection. ELISA has advantages of high sensitivity, low cost, easy to use, reliable and fast application over other techniques [46, 47]. ELISA can be used for the routine inspection of milk for the adulteration [48]. According to FSSAI, ELISA can be used as one of the techniques for melamine detection [49].

2.3. DNA Based Methods/Molecular Techniques

DNA based methods for adulteration detection is useful in certain cases like in traded commodities of plant origin, specifically, in case of detecting biological adulterants. Differentiation of adulterants from original food can be done by DNA/ Molecular-based techniques. Polymerase chain reaction (PCR), sequencing and hybridization are three major DNA based techniques employed for the detection of an adulterant.

Genomic content is very specific so any variation in specific gene segment due to deletion, insertion or transversion can give an idea about the adulterant and can be detected [50]. Most of the biological organism including the contaminants has some specific repeated sequences. Identification of these organisms can be done by amplifying the specific sequence using primers in PCR. This can alternatively be done by sequencing by hybridization methods, which can detect a small change in the nucleotide strand relative to the known DNA sequence [51].

Physical and chemical methods are good and easy to use for routine adulteration detection from food. However, these methods do not provide exact qualitative and quantitative data. The structural analysis requires high expertise in the analysis of micro and macroscopic features of plant substitutes in food [52]. Synthetic drug or phytochemical can be detected by a chemical profiling. Chemical profiling for an adulteration detection is still not be used because it contains chemical standards which may be expensive or rare, and in some cases, marker compound is unknown [20]. Although, molecular methods like hybridization and sequencing-based technique are desirable for biological adulterant detection. However, prior sequence knowledge is required for designing primers [53]. Moreover, a large amount of DNA is needed and these processes are time-consuming, labor-intensive and involve rigorous experimental conditions as compared to PCR based methods [54]

2.4. Nanotechnology-Based Methods

Old laboratory methods for detection of the toxins and chemical contaminants of the food are relatively sensitive and specific for the detection of contaminants and tested analysts, arguably, most of these tests are laborious, time-consuming and require skilled laboratory personnel [55, 56] which are unsuited for point-of-care testing.

Nanotechnology is a rapidly growing arena comprising nanomaterials. Its applications in the field of food safety range from the beginning of food processing to the end point of detecting contaminants in the food. Nanomaterials have been used to selectively remove contaminants [57], as antioxidants [58] and an antimicrobial [59]. Besides from sensors for detection of chemicals, heavy metals, pesticides, toxins, and foodborne pathogens nanotechnology is also used in extraction practices to adsorb trace amounts of insecticides [60, 61].

Nanotechnology-based detection methods have various advantages over conventional methodologies. Nanotechnology that uses nanoparticles in combination with electrochemical or optical detection modalities offers rapid, sensitive, and cost-effective techniques that permit miniaturization and automation for point-of-care testing. The unique properties of nano-devices and nanomaterials in combination with correct ligands offer unique opportunities to develop ultrasensitive detection strategies. Various nanomaterials have been used to identify chemicals, heavy metals and various toxins which include nanoparticles, spin valve nanosensors, nanotubes, nanowires, and quantum dots [62-65]. Nanotechnology-based detection methods differ in their mechanisms and designs but they share the same objective, that is, sensible and correct detection of trace pathogens or other contaminants [66].

2.4.1. Magnetic Nanoparticles

Magnetic nanoparticles (MNPs) are <100 nm in a size that can be influenced by an outer magnetic field. In clinical and medical applications MNPs have attracted a significant interest for years by unique advantages they comprise over conventional detection methods. Biological materials contain minor magnetic properties, and thus highly sensitive measurements can be done in visually obscured and minimally processed samples [67]. Magnetic nanoparticles are nontoxic, inexpensive to produce, biocompatible, environmentally safe and physically and chemically stable [68].

In very small MNPs, under the influence of thermal fluctuation magnetization can randomly flip polarity (superparamagnetism) the contaminant can be detected [69]. Superparamagnetic property occurs in a single domain MNPs of size less than or equal to tens of nanometers in diameter, depending on the material used [70]. These superparamagnetic nanoparticles (SPNs) display a negligible remnant magnetic moments in the absence of an external magnetic field, and their magnetic moments raise as the external field increases. This property prevents SPNs from aggregating in physiological solutions. Most commonly used MNP is superparamagnetic iron oxide nanoparticles (SPIONs), typically used for immunoassays by changing their surfaces with suitable ligands that can bind precisely to the target of interest.

2.4.2. Non-Magnetic Nanoparticles

2.4.2.1. Localized Surface Plasmon Resonance

Surface plasmon resonance (SPR) is a normally used optical detection technique for marker-free biological sensing. In this method, a polarized monochromatic light beam is passed through a prism and reflected from the thin metal film (typically gold) on a glass slide, which is in contact with the liquid solution of concern. The SPR is attained when the frequency of the incident photons striking the metal film matches with the normal frequency of the oscillating surface electrons on the surface of the metal. Meanwhile, the change of the resonance frequency is relative to the change in the surface concentration of species adsorbed on the metal surface, real-time binding capacities can be achieved by acquiring the change of optical reflectivity with time [71]. Even though SPR technology is well-established, the instrumentation engaged in this technique is expensive and difficult to operate [72].

To overcome these, localized Surface Plasmon Resonance (LSPR) has been developed. Here charge density oscillation can be excited by coupling to the excitation light due to their confinement in the subwavelength nanoparticles. Detection of pathogenic bacteria at ultra-low limits has been achieved with the help of LSPR [73, 74]. A copper-capped nanoparticle array with silica cores was presented whose plasmonic properties can be altered by changing the width of the copper shell [74]. These nanoparticles were functionalized with species-specific single-stranded DNA probes which were subsequently conjugated with target DNAs amplified by PCR. Different pathogenic bacteria such as *S. aureus* and *Enterococcus faecalis* were confirmed, with a limit of detection of 10 fM of DNA [75]. LSRP can further be improved by employing nanostructures with sharp features. Star-shaped gold nanoparticles were produced by Lee et al. [76]. It was found that multiple nanogaps can exist in each cluster of nanostars, which magnifies the signal via plasmon coupling. The nanostar clusters exhibited up to 460-fold higher field density than gold nanosphere clusters of similar mass. Nanoparticles with new shapes were also demonstrated, such as nano rice [77] and nanocrescents [78]

2.4.2.2. Nanoparticle-Assisted Colorimetry

Simplest detection techniques for biological targets are colorimetric detection. By using the chemical reaction resulting color change, the presence of target analyte can be determined either qualitatively by visualization or quantitatively using colorimeters. However, the efficacy of color conversion in conventional colorimetry is comparatively low, which reduce the sensitivity of the device. To overcome this problem, nanoparticle-assisted colorimetry was developed [79]. By introducing nanoparticles into the sensing matrix, the surface area on which color precursor molecules are placed can be improved significantly.

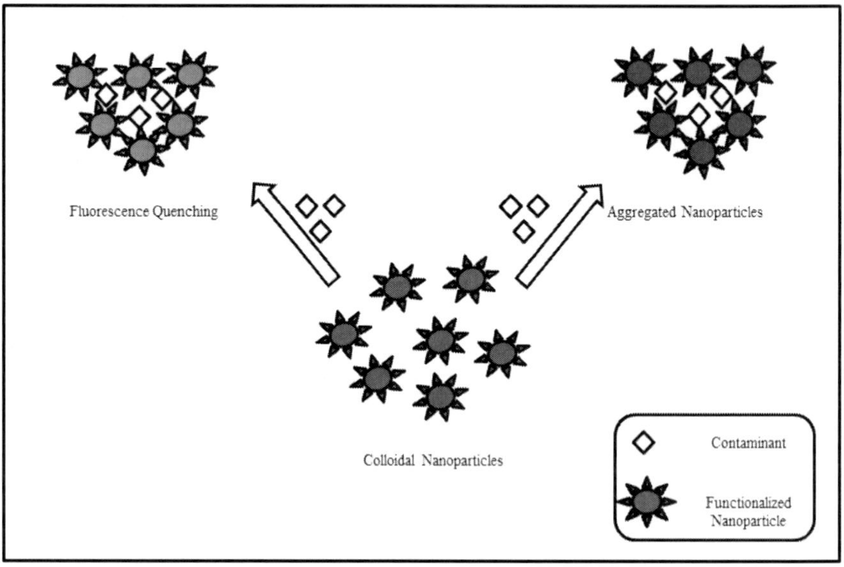

Figure 2. Schematic representation colorimetric and fluorometric change in colloidal nanoparticles due to aggregation in the presence of a contaminant.

Combining DNAzyme probe self-assembled gold nanoparticles and polymerase chain reaction (PCR) for the detection of the *invA* gene of *Salmonella* was demonstrated by Luo *et al.* [80]. A capture probe immobilized on a plate was used to form a sandwich structure, through hybridization between target sequence, capture probe, and a detection probe using functionalized AuNPs. In the presence of hemin, the detection probe could form a G-quadruplex/hemin complex, which aided as a catalyst for the oxidation reaction mediated by H_2O_2, leading to an intense color change. Detection limit was 3 x 10^3 CFU/mL. Similarly DNA detection, AuNPs has also been conjugated with antibodies for the detection of *Salmonella typhimurium* [81]. Melamine at a concentration of 10μM was detected using AMTr coated gold nanoparticles. Colorimetric change is due to the aggregation of nanoparticles in the presence of contaminant [82] (Figure 2).

2.4.2.3. Quantum Dots

Although extensively useful, traditional fluorescence detection with organic fluorescent dyes often shows overlapping emission peaks and high sensitivity to photobleaching. Quantum dots (QDs) which are highly resistant to photobleaching and chemical degradation are semiconductor nano-sized crystalline clusters [83]. The bandgap energy of QDs is inversely proportional to their size, which gives fine and tunable emission bands, independent of the excitation wavelength [84]. They are offered with various sizes ensuring a wide range of emission wavelength, which facilitates its use in the multiplex analysis.

Quantum dots, combined with an immunomagnetic separation (IMS), were used in multiplex detection of *E.coli* and *Salmonella enteritidis* [85]. Magnetic beads and QDs functionalized with corresponding antibodies and bound to target antigens were concentrated using a magnet. The fluorescent intensities produced by capturing different concentrations of bacteria were measured with a working range of 5×10^2 to 5×10^4 CFU/mL and 4×10^2 to 4×10^5 CFU/mL for *E. coli* and *S. enteritidis* respectively. QDs for real-time detection of foodborne bacteria in food materials has also been reported [84, 86]. The latest report demonstrated the detection of parathion, a pesticide, using functionalized graphene quantum dot (GQD) based electrochemical immunosensor with a detection limit of 46 pg/L [87].

Conclusion and Future Perspectives

Nanotechnology has enormous potential as a theranostic in the field of medicine. Also, nanotechnology has a various applications in the food industry for food safety, where it requires a real-time, simple, rapid and cost-effective method to detect contamination of foodborne pathogens, toxins, and chemicals. Nanotechnology can provide high-throughput and handy analytical platforms for screening pathogenic microbes and contaminants directly from food samples in the field and point-of-care clinical settings. Many of these technologies provide multiplexing abilities for real-time testing of multiple analytes with extremely high sensitivity that can considerably decrease the rate of testing and detection time. Although nanotechnology-based methods for detection of food contaminants have numerous advantages over traditional methods such as time, sensitivity, portability, and cost. The inherent complexity of food samples is challenging for the effective implementation of these new techniques in the food industry as a general sensor for different types of food samples. These technologies have been assessed with real food samples acquired from the field, but have only been tried with a selected type of foods. Assessment with an enormous and diverse range of food samples and contrast with well-established methods may help to address this challenge

Conflicts of Interest

There are no conflicts to declare.

Acknowledgments

Funding received from the Department of Biotechnology, Government of India under the project "NanoToF: Toxicological evaluation and risk assessment on Nanomaterials in Food" (grant number BT/PR10414/PFN/20/961/2014) and DST SERB Project "Nanosensors for the Detection of Food Adulterants and Contaminants" (grant number EMR/2016/005286) is gratefully acknowledged. Financial assistance by The Gujarat Institute for Chemical Technology (GICT) for the

Establishment of a Facility for environmental risk assessment of chemicals and nanomaterials is also acknowledged.

REFERENCES

[1] Division FaN. *Living well with HIV/AIDS.* http://www.fao.org/docrep/005/Y4168E/y4168e04.htm#P98_15090: FAO 2002.

[2] Ranga.nr. *Importance of food-in life & health* https://www.studyread.com/importance-of-food/: Study Read; 2017.

[3] Schell LM, Gallo MV, Cook K. What's NOT to eat—food adulteration in the context of human biology. *American Journal of Human Biology.* 2012;24(2):139-48.

[4] Robertson LJ, Sprong H, Ortega YR, van der Giessen JW, Fayer R. Impacts of globalization on foodborne parasites. *Trends in Parasitology.* 2014;30(1):37-52.

[5] Faille C, Cunault C, Dubois T, Bénézech T. Hygienic design of food processing lines to mitigate the risk of bacterial food contamination with respect to environmental concerns. *Innovative Food Science & Emerging Technologies.* 2018;46:65-73.

[6] Ingelfinger JR. Melamine and the global implications of food contamination. *New England Journal of Medicine.* 2008; 359(26):2745-8.

[7] Unnevehr LJ. Food safety issues and fresh food product exports from LDCs. *Agricultural Economics.* 2000;23(3):231-40.

[8] Mastovska K. Modern Analysis of Chemical Contaminants in Food. *Food Safety Magazine.* 2013.

[9] Rather IA, Koh WY, Paek WK, Lim J. The Sources of Chemical Contaminants in Food and Their Health Implications. *Frontiers in pharmacology.* 2017;8:830.

[10] Buculei A, Gutt G, Sonia A, Adriana D, Constantinescu G. Study regarding the tin and iron migration from metallic cans into foodstuff during storage. *Journal of Agroalimentary Processes and Technologies.* 2012;18(4):299-303.

[11] Cabado AG, Aldea S, Porro C, Ojea G, Lago J, Sobrado C, et al. Migration of BADGE (bisphenol A diglycidyl-ether) and BFDGE (bisphenol F diglycidyl-ether) in canned seafood. *Food and chemical toxicology*. 2008;46(5):1674-80.

[12] Lakshmi V. Food adulteration. *IJSIT*. 2012;1(2):106-13.

[13] Gould LH, Walsh KA, Vieira AR, Herman K, Williams IT, Hall AJ, et al. Surveillance for foodborne disease outbreaks-United States, 1998–2008. *Morbidity and Mortality Weekly Report: Surveillance Summaries*. 2013;62(2):1-34.

[14] Callejón RM, Rodríguez-Naranjo MI, Ubeda C, Hornedo-Ortega R, Garcia-Parrilla MC, Troncoso AM. Reported foodborne outbreaks due to fresh produce in the United States and European Union: trends and causes. *Foodborne pathogens and disease*. 2015;12(1):32-8.

[15] CDC. Surveillance for Foodborne Disease Outbreaks—the United States, 2009-2010. A*nn Emerg Med*. 2013;62:3.

[16] Song Q, Zheng Y-J, Xue Y, Sheng W-G, Zhao M-R. An evolutionary deep neural network for predicting morbidity of gastrointestinal infections by food contamination. *Neurocomputing*. 2017;226:16-22.

[17] Tirima S, Bartrem C, von Lindern I, von Braun M, Lind D, Anka SM, et al. Food contamination as a pathway for lead exposure in children during the 2010–2013 lead poisoning epidemic in Zamfara, Nigeria. *Journal of Environmental Sciences*. 2018;67:260-72.

[18] Sharma R, Agrawal M, Marshall F. Heavy metal contamination in vegetables grown in wastewater irrigated areas of Varanasi, India. *Bulletin of environmental contamination and toxicology*. 2006;77(2):312-8.

[19] Bhat R, Ramakrishna Y, Beedu S, Munshi K. Outbreak of trichothecene mycotoxicosis associated with consumption of mold-damaged wheat products in Kashmir Valley, India. *The Lancet*. 1989;333(8628):35-7.

[20] Shaw P-C, Wang J, But PP. *Authentication of Chinese medicinal materials by DNA technology: World Scientific;* 2002.

[21] Wilhelmsen EC. Adulteration determination. *Encyclopedia of Analytical Chemistry: Applications, Theory, and Instrumentation.* 2006.

[22] Bansal S, Singh A, Mangal M, Mangal AK, Kumar S. Food adulteration: Sources, health risks, and detection methods. *Critical reviews in food science and nutrition.* 2017;57(6):1174-89.

[23] Sheorey RR, Tiwari A. *Random amplified polymorphic DNA (RAPD) for identification of herbal materials and medicines–A review.* 2011.

[24] Mangal M, Bansal S, Sharma M. Macro and micromorphological characterization of different Aspergillus isolates. *Legume Res.* 2014;37(4):372-8.

[25] FSSAI. *Manual of Methods of Analysis of Foods (Spices and Condiments).* 2012.

[26] Louveaux J, Maurizio A, Vorwohl G. Methods of melissopalynology. *Bee world.* 1978;59(4):139-57.

[27] Ohe K. Scanning electron microscopic studies of pollen from apple varieties. *Acta Horticulturae (Netherlands).* 1991.

[28] Mouly P, Gaydou EM, Auffray A. Simultaneous separation of flavanone glycosides and polymethoxylated flavones in citrus juices using liquid chromatography. *Journal of chromatography A.* 1998;800(2):171-9.

[29] Blanch GP, Caja MdM, Ruiz del Castillo ML, Herraiz M. Comparison of different methods for the evaluation of the authenticity of olive oil and hazelnut oil. *Journal of agricultural and food chemistry. 1998;46(8):3153-7.*

[30] Silva B, Andrade P, Mendes G, Valentao P, Seabra R, Ferreira M. Analysis of phenolic compounds in the evaluation of commercial quince jam authenticity. *Journal of Agricultural and Food Chemistry.* 2000;48(7):2853-7.

[31] McDowell I, Taylor S, Gay C. The phenolic pigment composition of black tea liquors—part I: Predicting quality. *Journal of the Science of Food and Agriculture.* 1995;69(4):467-74.

[32] Calabrese M, Stancher B, Riccobon P. High-performance liquid chromatography determination of proline isomers in Italian wines. *Journal of the Science of Food and Agriculture.* 1995;69(3):361-6.

[33] Nogueira J, Nascimento A. Analytical characterization of Madeira wine. *Journal of agricultural and food chemistry.* 1999;47(2):566-75.

[34] Haughey SA, Graham SF, Cancouët E, Elliott CT. The application of near-infrared reflectance spectroscopy (NIRS) to detect melamine adulteration of soya bean meal. *Food Chemistry.* 2013;136(3-4):1557-61.

[35] Ozen BF, Mauer LJ. Detection of hazelnut oil adulteration using FT-IR spectroscopy. *Journal of agricultural and food chemistry.* 2002;50(14):3898-901.

[36] Nicolaou N, Xu Y, Goodacre R. Fourier transform infrared spectroscopy and multivariate analysis for the detection and quantification of different milk species. *Journal of Dairy Science.* 2010;93(12):5651-60.

[37] Wen Z-Q, Chen G, Luo Y, Li G, Masatani P, Cao X, et al. Adulteration Detection-Assessment of Spectroscopic Techniques for Adulteration Detection of Raw Materials Used in Biopharmaceutical Manufacturing. *American Pharmaceutical Review.* 2012;15(1):26.

[38] Alimentarious C. *Codex standard for named vegetable oils.* CODEX-STAN 210 2010; Amended 2003, 2005(8):1-13.

[39] Ruiz-Matute AI, Soria AC, Martínez-Castro I, Sanz ML. A new methodology based on GC− MS to detect honey adulteration with commercial syrups. *Journal of agricultural and food chemistry.* 2007;55(18):7264-9.

[40] Oliveira RC, Oliveira LS, Franca AS, Augusti R. Evaluation of the potential of SPME-GC-MS and chemometrics to detect adulteration of ground roasted coffee with roasted barley. *Journal of Food Composition and analysis.* 2009;22(3):257-61.

[41] Bogusz MJ, Hassan H, Al-Enazi E, Ibrahim Z, Al-Tufail M. Application of LC–ESI–MS–MS for detection of synthetic

adulterants in herbal remedies. *Journal of pharmaceutical and biomedical analysis.* 2006;41(2):554-64.

[42] FAO W. The international code of conduct on pesticide management. *World Health Organization and Food and Agriculture Organization of the …*; 2014.

[43] Souza EMTd, Arruda SF, Brandão PO, Siqueira EMdA. Electrophoretic analysis to detect and quantify additional whey in milk and dairy beverages. *Food Science and Technology.* 2000;20(3):314-7.

[44] Cartoni G, Coccioli F, Jasionowska R, Masci M. Determination of cows' milk in goats' milk and cheese by capillary electrophoresis of the whey protein fractions. *Journal of Chromatography A.* 1999;846(1-2):135-41.

[45] R. Vemireddy L, Archak S, Nagaraju J. Capillary electrophoresis is essential for microsatellite marker based detection and quantification of adulteration of Basmati rice (Oryza sativa). *Journal of agricultural and food chemistry.* 2007;55(20):8112-7.

[46] Bottero MT, Civera T, Anastasio A, Turi RM, Rosati S. Identification of cow's milk in "buffalo" cheese by a duplex polymerase chain reaction. *Journal of food protection.* 2002;65(2):362-6.

[47] Popelka P, Popelka P, Horská D, Golian J, Marcinčák S. Detection of sheep milk and cheeses adulteration using enzyme immunoanalysis (ELISA). *Slov Vet J.* 2002;27(3):36-7.

[48] Xue H-y, Hu W-w, Son H, Han Y, Yang Z-y. Indirect ELISA for detection and quantification of bovine milk in goat milk. *Food Sci.* 2010;31:370-3.

[49] India FSaSAo. *Advisory to the state health authorities on monitoring melamine contamination of foods.* 2008.

[50] Dhanya K, Sasikumar B. Molecular Marker Based Adulteration Detection in Traded Food and Agricultural Commodities of Plant Origin with Special Reference to Spices. *Current Trends in Biotechnology & Pharmacy.* 2010;4(1).

[51] Carles M, Cheung MKL, Moganti S, Dong TT, Tsim KW, Ip NY, et al. A DNA microarray for the authentication of toxic traditional Chinese medicinal plants. *Planta medica.* 2005;71(06):580-4.

[52] Joshi VC, Srinivas PV, Khan IA. Rapid and easy identification of Illicium verum Hook. f. and its adulterant Illicium anisatum Linn. by fluorescent microscopy and gas chromatography. *Journal of AOAC International.* 2005;88(3):703-6.

[53] Lockley A, Bardsley R. DNA-based methods for food authentication. Tren*ds in Foo*d Scienc*e & Technology.* 2000;11(2):67-77.

[54] Zammatteo N, Lockman L, Brasseur F, De Plaen E, Lurquin C, Lobert P-E, et al. DNA microarray to monitor the expression of MAGE-A genes. *Clinical chemistry.* 2002;48(1):25-34.

[55] Zhao X, Lin C-W, Wang J, Oh DH. Advances in rapid detection methods for foodborne pathogens. *J Microbiol Biotechnol.* 2014;24(3):297-312.

[56] Gracias KS, McKillip JL. A review of conventional detection and enumeration methods for pathogenic bacteria in food. *Canadian journal of microbiology*. 2004;50(11):883-90.

[57] Zhu J, Sadu R, Wei S, Chen DH, Haldolaarachchige N, Luo Z, et al. Magnetic graphene nanoplatelet composites toward arsenic removal. *ECS Journal of Solid State Science and Technology.* 2012;1(1): M1-M5.

[58] Deligiannakis Y, Sotiriou GA, Pratsinis SE. Antioxidant and antiradical SiO2 nanoparticles covalently functionalized with gallic acid. *ACS applied materials & interfaces.* 2012;4(12):6609-17.

[59] Kim JS, Kuk E, Yu KN, Kim J-H, Park SJ, Lee HJ, et al. Antimicrobial effects of silver nanoparticles. Nanomedicine: Nanotechnology, *Biology, and Medicine.* 2007;3(1):95-101.

[60] Yu X, Ang HC, Yang H, Zheng C, Zhang Y. Low temperature cleanup combined with magnetic nanoparticle extraction to determine pyrethroids residue in vegetables oils. *Food Control.* 2017;74:112-20.

[61] Savaliya R, Shah D, Singh R, Kumar A, Shanker R, Dhawan A, et al. Nanotechnology in disease diagnostic techniques. *Current drug metabolism.* 2015;16(8):645-61.

[62] Le Goff A, Holzinger M, Cosnier S. Enzymatic biosensors based on SWCNT-conducting polymer electrodes. *Analyst.* 2011;136(7):1279-87.

[63] Krishna VD, Wu K, Perez AM, Wang J-P. Giant magnetoresistance-based biosensor for detection of influenza A virus. *Frontiers in microbiology.* 2016;7:400.

[64] Lee H, Hwang NR, Hwang S-H, Cho Y. Magnetic nanowires for rapid and ultrasensitive isolation of DNA from cervical specimens for the detection of multiple human papillomaviruses genotypes. *Biosensors and Bioelectronics.* 2016;86:864-70.

[65] Liu J, Lau SK, Varma VA, Kairdolf BA, Nie S. Multiplexed detection and characterization of rare tumor cells in Hodgkin's lymphoma with multicolor quantum dots. *Analytical chemistry.* 2010;82(14):6237-43.

[66] Banerjee T, Shelby T, Santra S. *How can nanosensors detect bacterial contamination before it ever reaches the dinner table?* Future Medicine; 2017.

[67] Shao H, Min C, Issadore D, Liong M, Yoon T-J, Weissleder R, et al. Magnetic nanoparticles and microNMR for diagnostic applications. *Theranostics.* 2012;2(1):55.

[68] Wu K, Batra A, Jain S, Ye C, Liu J, Wang J-P. A simulation study on superparamagnetic nanoparticle based multi-tracer tracking. *Applied Physics Letters.* 2015;107(17):173701.

[69] Wu K. Magneto-Nanosensor Platforms for Biosensing Applications: University of Minnesota; 2017.

[70] Krishnan KM, Pakhomov AB, Bao Y, Blomqvist P, Chun Y, Gonzales M, et al. Nanomagnetism and spin electronics: materials, microstructure, and novel properties. *Journal of materials science.* 2006;41(3):793-815.

[71] Ferreira J, Santos MJ, Rahman MM, Brolo AG, Gordon R, Sinton D, et al. Attomolar protein detection using in-hole surface plasmon

resonance. *Journal of the American Chemical Society.* 2008;131(2):436-7.

[72] Chen S, Svedendahl M, Käll M, Gunnarsson L, Dmitriev A. Ultrahigh sensitivity made simple: nanoplasmonic label-free biosensing with an extremely low limit-of-detection for bacterial and cancer diagnostics. *Nanotechnology.* 2009;20(43):434015.

[73] Dasgupta N, Ranjan S, Chakraborty AR, Ramalingam C, Shanker R, Kumar A. Nanoagriculture and water quality management. *Nanoscience in Food and Agriculture 1:* Springer; 2016. p. 1-42.

[74] Kim D-K, Yoo SM, Park TJ, Yoshikawa H, Tamiya E, Park JY, et al. Plasmonic properties of the multispot copper-capped nanoparticle array chip and its application to optical biosensors for pathogen detection of multiplex DNAs. *Analytical chemistry.* 2011;83(16):6215-22.

[75] Fu J, Park B, Zhao Y. Limitation of a localized surface plasmon resonance sensor for Salmonella detection. *Sensors and actuators B: Chemical.* 2009;141(1):276-83.

[76] Park YI, Im H, Weissleder R, Lee H. Nanostar clustering improves the sensitivity of plasmonic assays. *Bioconjugate chemistry.* 2015;26(8):1470-4.

[77] Tanaka R, Yuhi T, Nagatani N, Endo T, Kerman K, Takamura Y, et al. A novel enhancement assay for immunochromatographic test strips using gold nanoparticles. *Analytical and bioanalytical chemistry.* 2006;385(8):1414-20.

[78] Unger A, Kreiter M. Analyzing the performance of plasmonic resonators for dielectric sensing. *The Journal of Physical Chemistry C.* 2009;113(28):12243-51.

[79] Ren J, Wang T, Wang E, Wang J. Versatile G-quadruplex-mediated strategies in label-free biosensors and logic systems. *Analyst.* 2015;140(8):2556-72.

[80] Luo R, Li Y, Lin X, Dong F, Zhang W, Yan L, et al. A colorimetric assay method for invA gene of Salmonella using DNAzyme probe self-assembled gold nanoparticles as a single tag. *Sensors and Actuators B: Chemical.* 2014;198:87-93.

[81] Wu W, Li J, Pan D, Li J, Song S, Rong M, et al. Gold nanoparticle-based enzyme-linked antibody-aptamer sandwich assay for detection of Salmonella Typhimurium. *ACS applied materials & interfaces.* 2014;6(19):16974-81.

[82] Vasimalai N, John SA. Picomolar melamine enhanced the fluorescence of gold nanoparticles: Spectrofluorimetric determination of melamine in milk and infant formulas using functionalized triazole capped goldnanoparticles. *Biosensors and Bioelectronics.* 2013;42:267-72.

[83] Bruchez M, Moronne M, Gin P, Weiss S, Alivisatos AP. Semiconductor nanocrystals as fluorescent biological labels. *Science.* 1998;281(5385):2013-6.

[84] Wang H, Li Y, Wang A, Slavik M. Rapid, sensitive, and simultaneous detection of three foodborne pathogens using magnetic nanobead–based immunoseparation and quantum dot–based multiplex immunoassay. *Journal of food protection.* 2011; 74(12):2039-47.

[85] Dudak FC, Boyaci IH. Multiplex detection of *Escherichia coli* and salmonella enteritidis by using quantum dot-labeled antibodies. *Journal of Rapid Methods & Automation in Microbiology.* 2009;17(3):315-27.

[86] Wang B, Wang Q, Cai Z, Ma M. Simultaneous, rapid and sensitive detection of three food-borne pathogenic bacteria using multicolor quantum dot probes based on multiplex fluoroimmunoassay in food samples. *LWT-Food Science and Technology.* 2015;61(2):368-76.

[87] Mehta J, Bhardwaj N, Bhardwaj SK, Tuteja SK, Vinayak P, Paul A, et al. Graphene quantum dot modified screen printed immunosensor for the determination of parathion. *Analytical biochemistry.* 2017;523:1-9.

In: Introduction of Forensic Nanotechnology … ISBN: 978-1-53616-040-6
Editors: R. K. Shukla and A. Pandya © 2019 Nova Science Publishers, Inc.

Chapter 8

ROLE OF NANOTECHNOLOGY IN FORENSIC DOCUMENT EXAMINATION AND PRESERVATION

Alok Pandya[1], Hiteshkumar B. Roz[2] and Ritesh Kumar Shukla[3,*]

[1]Department of Physical Science, Institute of Advanced Research, Gandhinagar, Gujarat, India

[2]Department of Law, Raksha Shakti University, Gandhinagar, Gujarat, India

[3]Biological and Life Sciences, School of Arts and Sciences, Ahmedabad University, Ahmedabad, Gujarat, India

ABSTRACT

Advent of nanotechnology plays a vital role in forensic investigation and examination. Forensic document examination is also not untouched with this technology. Nowadays, nanoparticles are used as security features (watermarks, fluorescent inks, security fibers, optical variable ink and holograms) into the documents, as security tags in passport, credit card and as a nano-tracker to monitor some suspicious elements. In

* Corresponding Author's E-mail: ritesh.shukla@ahduni.edu.in, shuklakritesh@gmail.com.

addition, nanoparticles are also used to restore and preserve the cultural heritage and old manuscripts. In this chapter an attempt was made to provide all the information regarding to application of nanotechnology in document examination and preservation, which may be helpful to open new avenue of research in to the field of forensic science.

Keywords: nanotechnology, forensic science, document examination, document preservation

1. INTRODUCTION

The integrity of the evidence is the most necessary part of forensic investigation and examination. In the case of forensic document evidences, it is very tricky because if the document gets destroyed then it is hard to revive. Keeping this in mind, forensic document examiner hence prefers non-destructive methods for the examination of the questioned document. In this line, the advent of nanotechnology plays a vital role in forensic investigation and examination. Forensic document examination is also not untouched with this technology. Nowadays, nanoparticles are used as security features (watermarks, fluorescent inks, security fibers, optically variable ink, and holograms) into the documents, as security tags in the passport, credit card and as a nano-tracker to monitor some suspicious elements [1]. In addition, recently, nanoparticles are also used to restore and preserve cultural heritage and old manuscripts [2]. The questions arise why we need to restore or preserve these things. The answer to this question is 'process of degradation.'

The degradation process is basically 2 types- 1-Natural degradation 2-Man-made degradation.

Natural degradation process arises from environmental factors, physical erosion and microorganisms, as well as handling and wrecking. On the other hand, the man made degradation process is intentionally done by a person with malicious intention. From a forensic point of view, man-made degradation processes always need to be investigated. In other words, alteration or destroy the questioned document is the kind of degradation process which can be done with harmful intention [2].

In this chapter, we made an attempt to provide all the information regarding the application of nanotechnology in document examination and preservation, which may be helpful for the forensic scientist, researcher, and students to open a new avenue of research into the field of forensic science.

2. Application of Nanotechnology in Document Examination

Document examination is very broad terms that include the examination of ink, paper, signature, handwriting, alteration (addition, deletion, obliteration) in written materials, etc. During all these examinations two things are very important which are as follows –

1. Integrity of the questioned documents (Preventive aspect)
2. Differentiation between genuine and counterfeit (Investigative aspect)

Keeping this in mind, nanotechnology plays a vital role in both preventive and investigative aspect.

In the preventive aspect, Nanotechnology is a unique resource to preserve the documents because nanomaterials do not alter the original physical and chemical properties of questioned documents and have a low environmental impact [2].

While, in the investigative aspect, in the form of various types of security features, nano-materials are intentionally added into the documents for security purposes.

2.1. Preventive Aspect

In the forensic document examination, the paper examination is a foremost step to proof whether the document is genuine or counterfeit. To ascertain the authenticity of the documents, forensic paper examination involves comparison and analysis of questioned and reference document. In many cases related to cheating, bank frauds, insurance, anonymous latter, wills copyright dispute, etc. the forensic document examiner encountered the problem of paper examination. At this point, nanotechnology can play its role by preserving the document for a long time and analyze it properly.

The conservation of historical papers started in 1933. Many methods and techniques were introduced to preserve paper and to reduce oxidation of ink of historical documents. William Barrow was the first person who introduced the field of conservation to paper. Barrow determined that instead of the origin of paper either rag or wood pulp, residues of sulfuric acid was the main factor of a rapid deterioration of paper. In 1966, Peter Waters who led a group of volunteers, called "mud angels," restored thousands of books and papers. It spurred resurgence in the profession of preservation and conservation worldwide. In 1987, Terry Saunders releases the film 'Slow Fires' on the preservation of the Human Record which examined ductility loss of paper resulting from acid decay.

According to Daniels et al. (1996), paper preservation can be divided into four areas [3]:

i. The chemistry of the deterioration of paper and the writing or images there upon.
ii. The physics and chemistry of processes that can be performed to enhance stability and appearance,
iii. The practical skills needed to carry out active conservation processes, e.g., Bleaching and Flashing, and
iv. The management and storage of a collection [4].

Baglioni and Rodorico et al. reported that the most recent applications of nanotechnology are in the field of conservation and restoration of the world's cultural heritage [5]. According to this study, Nanoparticles of soft calcium and magnesium hydroxide and carbonate can be used to restore and protect wall paints and to de-acidify paper and wood. In 2006, Baglioni group had reported that magnesium hydroxide nanoparticles suspended in alcohols can be used to inhibit two different and synergistic degradation processes usually affecting historically valuable manuscripts and paper documents [5].

To address preservation issues, it is necessary to find stable and well-dispersed nanoparticles in a non-aqueous solvent, because water can cause hydrophilic layers on the paper to swell and/or inks to leach out. This requirement prompted the application of nanoparticles that are stably dispersed in short-chain alcohols.

From these studies, two properties of nanomaterials came out for paper preservation [6]

1. Nanomaterials should have opposite chemical properties of ink composition present on paper.
2. Nanomaterials should have physical and chemical properties to inhibit the degradation process and minimize environmental impact.

Recently, Emma Sánchez at the Instituto del Patrimonio Cultural de España (IPCE), a special institute for the restoration of books and documents, revealed the new paradigm of nanotechnology in paper/document restoration. According to Emma, nanoparticles of same chemical composition ($Ca(OH)_2$) with nano-scale in size can easily penetrate among the paper fibers so that the chemical not only interacts more homogeneous and thorough with them but decrease the probability of creating whitish areas on the paper surface. Furthermore, it is more interesting because of its application to deacidification, which often blurs the intensity of black or brown inks. But, above all, because further, the product penetrates into the document, the more complete is the deacidifying effect. When the particle diameter is smaller, it fits in narrower spaces among the fibre weft and so its effect is deeper throughout the paper [7].

From these studies, one thing is very clear that nanotechnology has enormous potential to grow in the field of preservation/restoration of paper/documents. But same time various factors may also influence the process of preservation such as pH of the paper, the composition of ink present on the paper, type/composition of paper, the porosity of the paper, etc. Therefore, before doing paper preservation researcher/examiner should know about the characteristics of the paper and accordingly used/modify the technique.

2.2. Investigative Aspect

2.2.1. Formulation of Ink

From ancient times different manufacturers of ink have used various sets of colorants in ink formulation for their individualization. The colorants of ink are normally either pigments or dyes or both. The major components of the ink formulation are the solvents while the minor, but important components include surfactants, viscosity adjusters, preservatives, corrosion control agents, and driers. The nature of the colorants and solvents depends on the type of ink being made [8].

The forensic examination of inks consists of optical, physical, and chemical examinations. The results from the optical, physical and chemical examinations create the analytical profile of the ink [8]. This profile is compared with those of in a standard reference collection and when a match is found, one can infer who the manufacturer of the ink is, where it is sold, when the ink first became commercially available, in what pen or pens the ink is used (in the case of writing inks), and in what printers they are used (in the case of printing inks). NPs play an important role in all three types of examination of ink. It may help to make a profile of ink more unique and precise. In the recent development Ink jet inks have been introducing nano-sized pigments to their normally dye-based inks to give them added light-fastness. Nowadays, some fluorescence tagged NPs are also used in ink formulation [1].

Recently, various types of nanoparticles are developed that can be used into a new generation of anti-counterfeiting inks. These NPs are not only replacing the ancient fluorescent dyes that can be used in currency notes but also gives more precise and upgrade security features than previously [8].

2.2.2. Ink Analysis

Nanotechnology can play an important role in forensic ink examination. Nowadays, different ink manufacturer used nano-materials as one of the component of the ink especially gel and ink-jet inks [1]. Forensic ink examinations are mostly relying on chromatographic techniques but nano-materials cannot be characterized and identified by using these techniques. So to characterize this nano-materials elemental and morphological (size & shape) profiling scanning electron microscope coupled with energy dispersive X-ray (EDX) could be one of the most suitable technique. Addition of SEM-EDX along with existing chromatography techniques would provide better and unique profiling of ink analysis [1, 8].

2.2.3. Nano-Particles as Security Features

Security of official and confidential documents is always a big issue for a country regarding their national security. Therefore, time to time the ruling government of the countries is making or attempting to make a document secure and counterfeit-proof. These include watermarks, fluorescent inks, security fibers, optical variable ink, and holograms. There are various ways to secure the document. There are three main reasons for executing security features on objects of value (including documents) is for product labeling, document security, and object identification [9].

The uses of nano-sized particles in documents as security features are not new. These are used in most of the paper as fillers such as TiO_2, $CaCO_3$, and $BaSO_4$, all of which can be ground to submicron size. Also, most printing inks and whiteners are containing pigments which can also be ground to submicron size. Fillers and pigments are normally not part of a document's suite of security features. However, if one wants to mark (tag) a document with an invisible code that is luminescent, one normally uses inorganic luminescent phosphors or organic luminescent fluorophors.

Decrease the size of the inorganic phosphors to submicron size reduces their luminescence efficiency. This is where the attention in using nanoparticles as security features come in, particularly, luminescent nanoparticles such as quantum dots or nano-sized luminescent phosphors and up-converters. Thus quantum dots and other nanoparticles are being considered as replacing the more traditional micron-sized luminescent materials [1, 9].

2.2.4. Nanoparticles as Security Tags

NPs as security tags are also used in Passport and credit cards to upgrade in their security features. These nano-based security features cannot be easily counterfeit because due to nano in size cannot be visualized by naked eyes magnifying lens or any of the normally used techniques such as transmitted light, oblique light. It can be visualized only by the instruments which are highly precise and throughput (in magnification and various light sources) not found easily and present only in laboratories. Therefore, the chances of counterfeiting are much less than previously [1, 9].

2.2.5. Nano Trackers

Nowadays, trackers and barcodes are used to combat crime. Trackers are being used as a preventative tool in the form of secret patterns on the products to avoid people stealing. Apart from preventing tool, trackers can also help to track down those items which are stolen or missing. Trackers can also be used for security purpose as they can be used to avert prisoners from absconding. In this case, nano trackers are injected into the prisoners' body which helps to track them if they escape. Nano trackers are also enabling to monitor the convicts after being released. Nano trackers help to track them easily if they again involved in crime [10].

In addition, barcodes also help to prevent crime [11]. Nanotechnology helps in determining the authenticity of the items. Nanotechnology can also be applied in various sectors to prevent them from forged or duplicate products [11]. Counterfeiting or duplication of products leads to major economic loss and reputation of the respective companies and their countries. Product authenticity is directly linked to the reliability and reputation of the company, if the customer purchases the forged or duplicate products then reliability of the products and reputation of the company will be gone. If companies are fabricated forged products and pass them off as authentic may affect both consumers and companies as they suffer losses. Uses of watermarks may help to prevent duplication of products by differentiating genuine from forged products [11]. With the help of nanotechnology various tools are being developed that could be able to differentiate forged products from an authentic one. In this way, nanotechnology helps to prevent counterfeit crime. Nanotechnology has been developed nano-fibres and nano-dots with diverse colors to detect duplicate products. These nano-fibres and nano-dots have been detected using various forensic light sources that are meant to decipher the nano-fibres and nano-dots. So, in this way, counterfeit crime has been prevented [12].

Additionally, different types of criminal cases like hit and run, burglary and fraud can also be investigated with the help of nanotechnology [13]. Nanotechnology is also relevant in those cases in which petroleum products are involved such as arson cases to examine the petroleum hydrocarbon traces. Furthermore, nanotechnology is also pertinent in the cases related to cyber [mobile & internet] and sting operations [authentication of tapes and identification of speakers]. In these cases, the investigators scrutinize suspected tapes by identifying the coating of nanomaterials on its surface which is marking of authentication.

CONCLUSION

In conclusion, this chapter clearly demonstrated that there is plenty of scopes to develop new nanomaterials with unique physical and chemical properties to preserve the documents. Furthermore, nanotechnology can provide various methods to detect and identify the security features and differentiate authentic and counterfeit documents.

Although nanomaterials offer new opportunities for the preservation and examination of questioned documents, the environmental and health concerns surrounding their use must be critically addressed.

Future Perspectives

Nanotechnology can provide immense innovative approaches not only to preserve the questioned document but also help to develop new methods for document examination. But many 'if and but' may create obstacle in the field of documents preservation, hence so careful examination must be needed. Application of nanotechnology in preservation and examination of the document are still inadequate. Scarcity of available literature, methods and its optimization, various influential factors like pH of the paper, the composition of ink present on the paper, type/composition of paper, and porosity of the paper create hurdle in the research in this untouched area. Future forensic research should be focused on this thrust area of forensic science which indeed have the potential to shift the paradigm of the investigation process and open new challenging avenues of research.

References

[1] Shukla RK. Occupational exposure of nanoparticles in forensic science: a need of safe use. *Int. J. Fore. Sci. Pathol.* 2013;1(3):7-10.

[2] Baglioni P, Carretti E, Chelazzi D. Nanomaterials in art conservation. *Nature Nanotechnology*. 2015;10(4):287.

[3] Chelazzi D, Poggi G, Jaidar Y, Toccafondi N, Giorgi R, Baglioni P. Hydroxide nanoparticles for cultural heritage: consolidation and protection of wall paintings and carbonate materials. *Journal of colloid and interface science*. 2013;392:42-9.

[4] Daniels VD. The chemistry of paper conservation. *Chemical Society Reviews.* 1996;25(3):179-86.

[5] Baglioni P, Giorgi R. Soft and hard nanomaterials for restoration and conservation of cultural heritage. *Soft Matter.* 2006;2(4):293-303.

[6] Ambrosi M, Dei L, Giorgi R, Neto C, Baglioni P. *Colloidal particles of Ca (OH) 2: properties and applications to restoration of frescoes.* Langmuir. 2001;17(14):4251-5.

[7] http:// ritaudina.com/ en/ 2014/ 06/ 24/ nanotechnology-chemical-gels-applied-to-paper-restoration/).

[8] Campos-Cuerva C, Zieba M, Sebastian V, Martínez G, Sese J, Irusta S, et al. Screen-printed nanoparticles as anti-counterfeiting tags. *Nanotechnology.* 2016;27(9):095702.

[9] Pandya A, Shukla RK. New perspective of nanotechnology: role in preventive forensic. *Egyptian Journal of Forensic Sciences.* 2018;8(1):57.

[10] Hogarth E ER. *Community safety.* Chester Academic Press. 2007.

[11] Gooch J, Goh H, Daniel B, Abbate V, Frascione N. Monitoring criminal activity through invisible fluorescent "peptide coding" taggants. *Analytical chemistry.* 2016;88(8):4456-60.

[12] Teeuw WB, Vedder AH, Custers BH, Dorbeck-Jung BR, Faber EC, Iacob SM, et al. *Security applications for converging technologies*: Boom Juridische uitgevers; 2008.

[13] Chalmers JM, Edwards HG, Hargreaves MD. *Infrared and Raman spectroscopy in forensic science*: John Wiley & Sons; 2012.

In: Introduction of Forensic Nanotechnology ... ISBN: 978-1-53616-040-6
Editors: R. K. Shukla and A. Pandya © 2019 Nova Science Publishers, Inc.

Chapter 9

FORENSIC EVIDENCE: CASES, LEGAL ASPECTS AND ASPECTS OF NANOTECHNOLOGY

Heena Goswami [*] ***and Aathama Kumar***
Division of Science and Technology, Gujarat National Law University,
Gandhinagar, Gujarat, India

ABSTRACT

Nanotechnology is a powerful emergence of science and technology which can be used any stream of science including physical, chemical, biological, computational and even in the science of crime detection that is well known as forensic science. All technology has its own pros and cons but when it is integrated with forensic science then along with ethical issues, legal concern also rose and can be addressed. In this chapter, we attempted to elucidate the role of evidence in the justice delivery system and how legal proceedings can be governed with the assistance of newly developed technology in terms of its application and acceptability.

Keywords: nanotechnology, forensic science, law

---

[*] Corresponding Author's E-mail: hgoswami@gnlu.ac.in.

1. NARCOTICS

1.1. Introduction

The Narcotic Drugs and Psychotropic Substance Act (1), in Schedule I and II, encompass what comes under the ambit of Narcotics. In short, narcotics are drugs whose production, sale and purchase are prohibited, controlled or regulated. However Narcotic drugs are those which induce the human mind to an induced altered state of consciousness and have high abuse potential.

The Narcotic Drugs and Psychotropic Substance Act and the Prevention of Illicit Traffic in Narcotic Drugs and Psychotropic Substance Act, 1988(2)are legislations that govern the same in India. Although the Indian Evidence Act is separate legislation in itself it also has overlapping provisions and its comparisons which will also be dealt with in this chapter.

1.2. History: An Overview of Drug Control Laws in India

Atharvaveda has its first mention of Cannabis smoking that proves that India is a country where Cannabis smoking was prevalent and not seen as a wrong act (3) In the year 1952 under the British Era, the Opium Act (4) was brought into force and further The Opium Act of 1878 (5) and finally the Dangerous Drugs Act, 1930 (6).

Although these provisions existed for the governance this turned out to be inadequate in the field of drug abuse at national and international level. Then in the year 1985, in order to bring India's Narcotics laws on par with International standards, the Rajiv Gandhi Government enacted the NDPS Act, banning all narcotic drugs in India including Cannabis which was till then legal. Thus the new Act grappled two kinds of offences: trafficking of prohibited substances i.e., cultivation, manufacture, distribution, and sale, as well as their consumption (7).

1.3. Legal Definitions

Narcotic Drugs are defined in Section 2 of the Act. Other relevant subsections include:

1.3.1. Cannabis

Cannabis-(a) Charas, that is the separated resin, in whatever form, whether crude or purified, obtained from the cannabis plant and also includes concentrated Preparation and resin known as hashish oil or liquid hashish (1); (b) Ganja, that is, the flowering or fruiting tops of the cannabis plant (excluding the seeds and leaves when not accompanied by the tops), by whatever name they may be known or designated (9); (c) Any mixture, with or without any neutral material, of any of the above forms of cannabis or a drink prepared therefrom (10).

1.3.2. Cannabis Plant

Cannabis plant means any plant of the genus cannabis.

1.3.3. Coca Derivative

Coca derivative means: (a) Crude cocaine, that is, any extract of coca leaf which can be used, directly or indirectly, for the manufacture of cocaine; (b) Ecgonine and all the derivatives of ecgonine from which it can be recovered; (c) Cocaine, that is, methyl ester of benzoyl-ecgonine and its salts; and (d) All preparations containing more than 0.1 percent of cocaine.

1.3.4. Coca Leaf

Coca leaf means:(a) The leaf of the coca plant except for a leaf from which all ecgonine cocaine and any other ecgonine alkaloids have been removed; (b) Any mixture thereof with or without any neutral material but does not include any preparation containing not more than 0.1 percent of cocaine; (c) "Coca plant " means the plant of any species of the genus Erythroxylum.

1.3.5. Controlled Substance

Controlled substance means any substance which the Central Government may have regard to the available information as to its possible use in the production or manufacture of narcotic drugs or psychotropic substances or to the provisions of any International Convention, by notification in the Official Gazette, declare to be a controlled substance.

1.3.6. Manufactured Drug

Manufactured drug means: (a) All coca derivatives, medicinal cannabis, opium derivative, and poppy straw concentrate; (b) Any other narcotic substance or preparation which the Central Government may, having regard to the available information as to its nature or to a decision, if any, under any International Convention, by notification in the Official Gazette, declare to be a manufactured drug; but does not include any narcotic substance or preparation which the Central Government may, having regard to the available information as to its nature or to a decision, if any, under any International Convention, by notification in the Official Gazette, declare not to be a manufactured drug.

1.3.7. Medicinal Cannabis

Medicinal cannabis that is, medicinal hemp, means any extract or tincture of cannabis (hemp).

1.3.8. Narcotic Drug

Narcotic drug means coca leaf, cannabis (hemp), opium, poppy straw and includes all manufactured drugs.

1.3.9. Opium and Opium Derivatives

1.3.9.1. Opium

Opium means (a) The coagulated juice of the opium poppy; and (b) Any mixture, with or without any neutral material, of the coagulated juice of the opium poppy, but does not include any preparation containing not more than 0.2 percent of morphine.

1.3.9.2. Opium Derivative

Opium derivative means: (a) Medicinal opium, that is, opium which has undergone the processes necessary to adapt it for medicinal use in accordance with the requirements of the Indian Pharmacopoeia or any other pharmacopoeia notified in this behalf by the Central Government, whether in powder form or granulated or otherwise or mixed with neutral materials; (b) Prepared opium, that is, any product of opium obtained by any series of operations designed to transform opium into an extract suitable for smoking and the dross or other residue remaining after opium is smoked; (c) Phenanthrene alkaloids, namely, morphine, codeine, thebaine and their salts; (d) Diacetylmorphine, that is, the alkaloid also known as diamorphine or heroin and its salts; and (e) All preparations containing more than 0.2 percent of morphine or containing any diacetylmorphine.

1.3.10. Opium Poppy

Opium poppy means: (a) The plant of the species Papaversomniferum L.; and (b) The plant of any other species of Papaver from which opium or any phenanthrene alkaloid can be extracted and which the Central Government may, by notification in the Official Gazette, declare to be opium poppy for the purposes of this Act.

1.3.11. Poppy Straw

Poppy straw means all parts (except the seeds) of the opium poppy after harvesting whether in their original form or cut, crushed or powdered and whether or not juice has been extracted therefrom.

1.3.12. Poppy Straw Concentrate

Poppy straw concentrate means the material arising when poppy straw has entered into a process for the concentration of its alkaloids.

1.3.13. Psychotropic Substance

Psychotropic substance means any substance, natural or synthetic, or any natural material or any salt or preparation of such substance or material included in the list of psychotropic substances specified in the Schedule (11).

1.4. Case Laws

1.4.1. Arif Khan v. State of Uttarakhand

The Supreme Court bench of Justices RK Agrawal and Abhay Manohar Sapre, in this case, held that Section 50 (12) of Narcotic Drugs and Psychotropic Substances Act, 1985 (NDPS Act) are mandatory and not a directory in nature. They also held that the searching and recovering of contraband should be in the presence of a Gazetted Officer or a Magistrate (13).

1.4.2. Union of India v. Mohanlal

The Supreme Court bench consisting of the Chief Justice T.S. Thakur and Justice Kurian Joseph in order to ‘avoid any confusion arising out of the continued presence of two notifications on the same subject has issued guidelines on the disposal of Narcotic Drugs and Psychotropic and Controlled Substances and Conveyance (14).

1.4.3. Mohinder Singh v. State of Punjab

The Supreme Court bench of Justices L. Nageswara Rao and Navin Sinha, on Thursday, in a case under the NDPS Act held that because NDPS act being presumptive in nature, the interpretation of the sections should be strict. The bench then restored the acquittal by the Additional Sessions Judge, Ludhiana of an accused, which was reversed by the Punjab and Haryana High Court (15).

1.4.4. Kishan Chand v. State of Haryana

Under, Section 42 (16) a mandatory pre-search requirement is that of recording whatever information received and sending it at once to the superior officer (17).

1.4.5. Sukhdev Singh v. State of Haryana

In this case, Search and seizure was conducted by the Investigation officer who later did not inform his higher authorities in writing instantaneously or after a reasonable delay. Here there is was evidence to provide that the delay was reasonable. It was held by the Supreme Court that there is a mandatory requirement to reduce the information in writing and send it forthwith to a superior officer. In case of any delay, it can only be in case of any special reasons. Accused was held to be acquitted (18).

1.4.6. Myla Venkateswarlu v. State of A.P

Under Section 50 (19) of the NDPS Act, there is a mandatory requirement of informing accused of the right to be searched before a magistrate or a gazetted officer. This needs to be strictly complied with (20).

1.4.7. Joseph Fernandez v. State of Goa

Under Section 50 of the act, a mere inquiry by said Investigating officer of whether the suspect

wants to be searched in the presence of a Magistrate or a Gazetted Officer cannot be said that the section has been complied with (21).

1.4.8. Raju Premji v. State of Punjab

In case of serious doubts as to the possession of contraband by the accused, the conviction was set aside. In this case, large quantity of heroin was seized by the officer in charge but the physical evidence to the recovery was not produced in the court. Furthermore evidence relating to the three samples taken from the bulk was also not produced. As this creates serious doubts, the conviction was set aside (22).

1.4.9. State of Rajasthan v. Daul Alias Daulat Giri

Under Section 8 (23) of the NDPS Act, on the sole reason that an employee in the office of S. P who was after the seizure in possession of the samples for a few hours and he was not seized, it cannot be the ground for acquittal (24).

In this case, Opium was seized from the possession of accused. All the requisites were fulfilled including seized material properly sealed and kept in custody (25) Samples had also reached FSL with seals intact. Here conviction was not set aside because the employee had not been examined.

1.4.10. Abdul Rashid v. State of Bihar

Under Section 8 of the NDPS Act, the Confessional statement made by accused to Superintendent of Excise under provisions of Bihar and Orissa Excise Act was Inadmissible in evidence as an excise officer is also a police officer under S.25 of Evidence Act (26). Here the conviction was based only on fact that accused was found together with the co-accused from whom the article was recovered. In this case, it was liable to be set aside (27).

1.4.11. Balbir Kaur v. State of Punjab

Under Section 15 (28) of the NDPS Act, it was held that mere delay in sending the sample to the chemical examiner is not a violation of Section 50, 52 and 57 of the NDPS Act. As long as the allegation of bias is not established, the search and recovery are not held to be vitiated in any manner and in this case on that basis, the conviction was set to be proper (29).

1.4.12. Krishan Mohar Singh Dugal v. State of Goa

In this case, the Charas was recovered with the help of the disclosure statement made by the accused. Evidence showed that the police were already informed about the place where charas was kept that is, Coconut tree from whose stem charas was found was standing on an open space accessible to all. It cannot, therefore, be said that it was accused who had concealed the charas there and that it was found out only on the basis of the disclosure statement made by accused Conviction of accused liable to be set aside (30).

1.4.13. Valsala v. State of Kerala

Under Section 21 the conviction was set aside of the fact that delay of over three months in sending the seized material to the court authorities and lack of evidence to show that the article was in proper custody and sealed properly made the seized article doubtful (31).

1.5. Critical Analysis

One can see how strictly the section of the NDPS Act is interpreted as far as Narcotics substance is considered. The high punishment given to the offenders put a huge weight on the judges that justice shall be served.

The Act has been criticized a number of times for its sharp contrast to the fundamentals of Criminal Law in India. Under Section 35, the Act presupposes the guilt of an accused and here the onus is on his to prove that he isn't. Furthermore, it is said that unless the contrary is proved, the accused shall be believed to have intentionally held the drugs that were found in his possession. Innocent until proven guilty, the very roots of this notion on which most of the statutes in India rests is violated by this section.

It is also widely believed that the act does not make a distinction between serious crimes and otherwise. No distinctions have been made between drug user, a hard addict, a petty peddler, and a seasoned drug trafficker and between hard and soft drugs. As to this particular reason many users are seen to move on to hard drug usage as the punishment is almost the same for both.

There have been several reports as to how people have exploited the new law to the fullest extent possible and have formed a close nexus with law enforcement agencies (32). One can ask the question here as to whether here Law has created more problems than it has solved.

2. Documents

2.1. Introduction

In the case of wrongful confinement by the police, the victim scratched his signature on one wall of the cell where he was detained illegally. When he came out, he filed a suit against wrongful confinement. The police denied the charge. The confinement was proved by his signature on the wall (33).

Documents play an important role in human life. Daily we sign a number of documents. Without signing or accepting or writing any documents, the entire world will be in chaos. In the example given above by B. R Sharma in his book Forensic Science in Criminal Investigation and Trials has just showcased how an engraving helped a victim to win his case.

A document is a material, having a symbol or writing on it, which conveys some meaning to one or more persons. Here it may be anything, from a piece of paper to a wooden piece or a piece of cloth (34).

2.2. Types of Documents

a) Letters
b) Financial Documents
c) Records
d) Tickets
e) Wills
f) Historical Documents
g) Books
h) Agreements
i) Identification Cards

2.3. Legal Definitions

"Document".—The word "document" denotes any matter expressed or described upon any substance by means of letters, figures, or marks, or by more than one of those means, intended to be used, or which may be used, as evidence of that matter.

2.3.1. Explanation 1

It is immaterial by what means or upon what substance the letters, figures or marks are formed, or whether the evidence is intended for, or may be used in, a Court of Justice, or not. Illustrations A writing expressing the terms of a contract, which may be used as evidence of the contract, is a document. A cheque upon a banker is a document. A power-of-attorney is a document. A map or plan which is intended to be used or which may be used as evidence is a document. A writing containing directions or instructions is a document.

2.3.2. Explanation 2

Whatever is expressed by means of letters, fig-ures or marks as explained by mercantile or other usage shall be deemed to be expressed by such letters, figures or marks within the meaning of this section, although the same may not be actual-ly expressed. Illustration A writes his name on the back of a bill of exchange payable to his order. The meaning of the endorsement, as explained by mer-cantile usage, is that the bill is to be paid to the holder. The endorsement is a document and must be construed in the same manner as if the words "pay to the holder" or words to that effect had been written over the signature (35).

2.3.3. Making a False Document

A person is said to make a false document or false electronic record—First —who dishonestly or fraudulently—

(a) makes, signs, seals or executes a document or part of a document;

(b) makes or transmits any electronic record or part of any electronic record;
(c) affixes any [electronic signature] on any electronic record;
(d) makes any mark denoting the execution of a document or the authenticity of the [electronic signature], with the intention of causing it to be believed that such document or part of document, electronic record or [electronic signature] was made, signed, sealed, executed, transmitted or affixed by or by the authority of a person by whom or by whose authority he knows that it was not made, signed, sealed, executed or affixed; or Secondly —Who, without lawful authority, dishonestly or fraudu-lently, by cancellation or otherwise, alters a document or an electronic record in any material part thereof, after it has been made, executed or affixed with [electronic signature] either by himself or by any other person, whether such person be living or dead at the time of such alteration; or Thirdly —Who dishonestly or fraudulently causes any person to sign, seal, execute or alter a document or an electronic record or to affix his [electronic signature] on any electronic record knowing that such person by reason of unsoundness of mind or intoxication cannot, or that by reason of deception practised upon him, he does not know the contents of the document or electronic record or the nature of the alteration (36).

2.4. Case Laws

2.4.1. Identification of Handwriting

2.4.1.1. Ram Chandra v. State of U.P.

It is unsafe to treat expert handwriting opinion as sufficient basis for conviction, but it may be relied upon when supported by other items of internal and external evidence (37).

2.4.1.2. Ishwari Prasad Mishra v. Md. Isa

In this case, it was held that the expert evidence of handwriting can never be decisive because it is, after all, opinion evidence and the conviction cannot be only based on this opinion (38).

2.4.1.3. Shashi Kumar Banerjee v. Subodh Kumar Banerjee

It was held by the Supreme Court that the expert evidence as to handwriting being opinion evidence can rarely if ever, take the place of substantive evidence and before acting on such evidence, it would be desirable to consider whether it is corroborated either by clear direct evidence or by circumstantial evidence (39).

2.4.1.4. Fakhruddin v. State of M.P.

This Court had again considered the evidentiary value of expert opinion in regard to handwriting and it uttered a note of caution pointing out that it would be risky to found a conviction solely on the evidence of a handwriting expert and before acting upon such evidence, the court must always try to see whether it is corroborated by other evidence, direct or circumstantial. It is interesting to note that the same view is also echoed in the judgments of English and American courts Vide Gurney v. Langlands (40).

2.4.1.5. Prasad Mahto v. Mt. Jasoda Koer

In this case, it was held that the court is competent to use its own eyes for the very purpose of deciding whether particular handwriting is similar or not. In a matter of this kind, the court cannot certainly act as automation accepting without scrutiny an expert's opinion as infallible (41).

2.4.1.6. Bhupendra Nath vs Union of India

The main point for determination in this reference is, whether an order directing an accused person to furnish his finger impressions for the purpose of comparison with certain disputed similar impressions under Section 73 of the Evidence Act or otherwise, would be contrary to the provisions of Article 20(3) of the Constitution. It was held in this case that it does not. Asking her handwriting or signature or finger impression for the purpose of comparison does not fall under the ambit of self-discrimination under Art. 20(3) (42).

2.4.2. Photographs

2.4.2.1. Laxmipat Choraria and Ors v. State of Maharashtra

If the court is satisfied that there is no trick photography and the photograph is above suspicion, the photograph can be received in evidence. It is, of course, always admissible to prove the contents of the document but subject to the safeguards indicated to prove the authorship. This is all the more so in India under s. 10 of the Evidence Act. To prove participation in a conspiracy. Detection and proof of crime will be rendered not only not easy but sometimes impossible if conspirators begin to correspond through photographs of letters instead of originals. But evidence of photographs to prove writing or handwriting can only be received if the original cannot be obtained and the photographic reproduction is faithful and not faked or false (43). In the present case, no such suggestion exists and the originals having been suppressed by the accused were not available. The evidence of photographs as to the contents and as to handwriting was receivable.

2.4.3. Newspapers

2.4.3.1. Ratan Lal Soni vs the State of Rajasthan and Ors

Newspaper is not one of the documents referred to in Section 78(2) of the Evidence Act by which an allegation of fact can be proved. The presumption of genuineness attached under Section 81 of the Evidence Act to a newspaper report cannot be treated as proved of the facts reported therein. A statement of fact contained in a newspaper is merely hearsay and therefore inadmissible in evidence in the absence of the maker of the statement appearing in court and deposing to have perceived the fact reported (44).

2.4.3.1. Laxmi RajShetty v. State of Tamil Nadu

The Supreme Court held that the newspaper is not admissible in evidence. A newspaper is not one of the documents referred to in Section 78(2) of the Evidence Act by which an allegation of fact can be proved (45).

2.4.4. Typewriting Evidence

2.4.4.5. State of Delhi v. S.J Choudhary

The reference made in the appeal to the Constitution Bench is for deciding the important question of law of whether the opinion of a typewriter expert is admissible in evidence under S.45 of the Indian Evidence Act. It was held that the opinion of a typewriter expert is admissible in evidence under S.45 of the Indian Evidence Act and the Apex Court allowed the appeal and seat aside what the lower courts held (46).

2.4.5. Tape Record

2.4.5.1. Ziyauddin Burhanuddin Bukhari v. Brijmohan Ramdass Mehra

Tape records of speeches are documents which stand on no different footing than photographs. A party who produces tape record of phone conversation must through witnesses prove the time, place and accuracy of the said tape recordings (47).

2.5. Critical Analysis

The falsification of financial documents is big business in almost every country and it is on the increase everywhere. It is estimated that hundreds of crores of rupees are lost in financial frauds.

The courts have always wary of accepting the findings of the experts as final and sufficient but there have been numerous times where documents have helped the victim from proving his case. At least documents help as corroborative evidence. Thus Documents always plays an important role in modern society to prove anything and everything.

3. Explosives

3.1. Introduction

The number of cases reported under Explosives and Explosive Substances Act in India was 4368 during the year 2015. The top 10 states in terms of the number of cases reported under Explosives and Explosive Substances Act were: Rajasthan, Tamil Nadu, Uttar Pradesh, Madhya Pradesh, Kerala, Chhattisgarh, Andhra Pradesh, West Bengal, Telangana, and Odisha during the year 2015 (48).

The number of cases reported under Explosives and Explosive Substances Act in Rajasthan was 824 during the year 2015. It accounted for 18.86% of the number of cases reported under Explosives and Explosive Substances Act in India during the year 2015 (49). This shows us that in India explosives are still used in any sabotage or political, communal riots, terrorist activities, dacoities, intimidation, suicides, burglaries. Even though explosives have a number of legitimate use to it like Blasting rocks for the construction of roads, railways lines, dams, tunnels or for mining works, firework display, etc. the abuse of them is in the rise.

This is mainly because of the easy improvisation that anybody can do and the easy availability of the materials and them being effective in all sorts of crime that includes even Political feuds. Explosives cause a sense of panic and fear among the victims which is taken advantage of by the offenders.

3.2. History

Before the advent of the Explosive Substances Act, 1908, the Explosives Act 1884 made provisions to regulate manufacturing, possession, using and selling of the explosive an offence. The Act also had provisions with giving license to the persons competent and requirements to get the license also fell under the provisions of the Act.

The Act also had prohibited any engagement of persons under the age of 18, and sentenced or convicted for an offence involving moral turpitude, persons against whom execution proceedings or order is passed by Civil Courts or even the persons whose license under the provision of the Act of 1884 was cancelled by the competent authority (50).

3.3. Legal Definitions

The Explosive Substances Act, 1908 (Act No. 6 of 1908) dated 8th June 1908 was enacted with an intention to amend the older laws. This act, with respect to section 1 of the act specifies the application of the act over any Indian and the jurisdiction extends to any Indian irrespective of his location inside and outside India. The act also extends its jurisdiction to the whole and entire India.

As defines by the act, Explosive substance shall be deemed to include any materials for making any explosive substance; also any apparatus, machine, implement or material used, or intended to be used, or adapted for causing, or aiding in causing, any explosion in or with any explosive substance; also any part of any such apparatus, machine or implement (51). The definition is quite wide in order to include any apparatus, machine, implement or material used or intended to be used, or adapted for causing, or aiding in causing, any explosion in explosive substance and also any part thereof. The Act was later amended in the year 2001 by The Explosive Substances (Amendment) Act, 2001, where the definition of Explosive Substances was further widened and the scope was further enlarged by including Special Category of Explosive Substance inclusive of the list of substances provided under clause (b) of Section 2 of the Act. To what extent, materials can be included under this particular act is not directly described in the act. But over the years this has been cleared by the Judiciary over different case Laws.

3.4. Case Laws

3.4.1. Masha Angami and Anr. vs Government of Manipur (1967)

In this, the Hon'ble Guwahati High Court held Gun Powder came under the ambit of an explosive substance under Section 2 of the Explosive Substances Act, 1908.

Further, the Court also observed that, if there is a prima facie reasonable suspicion that the person who is accused found in possession of gunpowder with the unlawful object then bail will be rejected (52).

3.4.2. Rajani Kanta Mandal v. State of Bihar

In this case, the apex court held that to form a charge under Section 5 of the Act, and to prove conscious possession of an Accused thereof is not sufficient, but such possession must be coupled with the unlawful object. Here the judiciary had tried and played an important role in the determination of the scope of the act (53).

3.4.3. Chandra Prakash v. State of Rajasthan

The question here was whether Ammonium nitrate not being an explosive substance, mere possession cannot bring the accused within the ambit of any offence. In this regard, the court referred to Section 4(d) of the 1884 Act.

The court also referred to Section 2 of the 1908 Act and held that it has a deeming provision which states that explosive substance would include any materials for making any explosive substance. Similarly, Section 4(d) of the 1884 Act has a broader spectrum which includes colored fires or any other substances, whether single chemical compound or a mixture of substances. That apart, the court found, apart from ammonium nitrate other articles had been seized. The combination of the same, as per the evidence of the expert witness, was sufficient to prepare a bomb for the purpose of an explosion. In addition to the same, a huge quantity of ammonium nitrate was seized and it was seized along with other items. The cumulative effect is that the possession of these articles in such a large quantity by the accused gives credence to the prosecution version that the possession was conscious and it was intended to be used for the purpose of the blast (54).

3.4.4. Giri Raj v. State

An explosive substance' can only be given the meaning as defined under the Explosive Substances Act, 1908 and the Cylinder as such, apparently does not fall within the definition of explosive substances (55).

3.4.5. Public Prosecutor, High Court of A. P. Hyderabad v. Dorasala Patnam Nadipi Veera Reddy

Explosive substances act without the due sanction of the government. But to obtain such sanction order it takes time and if the charge sheet... is filled without sanction order under the explosive substances act that by itself is not a ground to return the charge – sheet (56).

3.4.6. Mohammad Usman Mohammad Hussain Maniyar and Others v. State of Maharashtra

According to him the trial court should have framed charge against the respondents No. 2 and 3 for the offence Under sections 3 and 5 of the explosive substance act and not Under section 9b of the explosive act because huge quantity of explosive substance was found and in support of his submission he relied upon the judgment of the Hon'ble Supreme Court (57).

3.4.7. Nema Adak v. State of West Bengal

Here, in this case, there was no evidence to prove that any injury not to speak of serious injury was caused by bombs and not by crackers, but the fact alone would not satisfy the requirements of section 3 under the Act (58).

3.4.8. Balachandra v. State of Maharashtra

In this case the Supreme Court held the accused liable under the act not on the basis on direct evidence of the immediate cause of the explosion but indisputably the explosives the possession of which was prohibited by the act was found in the shops of the accused or the premises where the appellants carried their business and the substances that were carried out of very hazardous nature (59). The accused should have completely avoided the use of highly sensitive compositions of the nature mentioned above.

3.5. Critical Analysis

With the number of explosives used for different types of violence in India, the act seems to have a positive effect in spite of it being one of the oldest acts. Different amendments and rules have come up to exactly underline the interpretation of the present act.

4. Law Enforcement Use of Nanotechnology

Nanotechnology promises smaller and more accurate tools and systems, batteries with longer life and extremely sensitive sensors. The nanotags (The Indian Police Journal, 2007) will be used to coat surfaces to capture DNA, nanoparticles will develop miniature sensors that will detect different illegal drugs, x-ray systems will be improved through the use of carbon nanotubes, infrared nanotechnology will expand the uses of searching for missing people and suspects, dust (known as Clever Dust, Deputy Dust, Daft Dust, Smart Dust, and Flashy Dust) may be used to track movements, and miniscule Radio Frequency Identification may offer assistance to those who need them the most. (UH ISRC, 2005). The pledge of using nanotechnology to investigate criminal activity is rapidly resembling and the use of these newly developing tools may provide the next generation of peace officers their primary method to solve crimes. According to Lieutenant Colonel Nate Allen, U.S. Army, "the pace of change mandates that efficient leaders understand the insinuations and promise that the coming technological tsunami holds." (Halal, 2008) Daniel Linstedt, an internationally known expert in data warehousing business intelligence, proclaimed "Nanotech is already here, and there are infinite utilizations for it." (Linstedt, 2010) Mike Pitlethy, the CEO of ROAR Particles, concurs. In 2009, Pilethy remarks on law enforcement agencies have begun to evaluate and use nanotech tools. For occasion, the FBI started using a nanoscale developer and x-ray source to image engraved fingerprints on a number of high profile cases (Pitlethy 2009). This type of nanotechnology developed by researchers at Leicester University in England assists with the detection of fingerprints left on a bullet casing which can capture a print after the chemicals from the print corrode the surface of the metal. Nanoscale powders provide clearer images and their prints developed using nanoscale powders can provide significant information about whether or not the donor had been handling explosives, illegal drugs or if the donor had been using illegal drugs. Even though It is difficult to imagining how this could be used, a recent incident receiving national attention exemplifies the perspective of this breakthrough. In New York's, May 1, 2010, Times Square, a suspected terrorist abandoned a suspected Vehicle Borne Improvised Explosive Device (VBIED).The investigating law enforcement agency could have ascertained explosive compounds while developing leads from fingerprints. Using nanotechnology, explosive detection would have possible the investigator's valuable officer safety information and excellent trial evidence. Auspiciously, the police made a quick arrest without any harm officers or the community. Near future, information obtained through explosive detection nano-tools will better ensure the safety of law enforcement officers. Soon, it may be possible to avert VBIED and similar events altogether if the appropriate surveillance nano-tools are deployed in places like New York and other terrorist targets. The future promises other prevailing tools that will prevent crime, safely lead investigators to perpetrators and instantly locate missing at-risk people. Currently, the U.S. Army is testing a wireless health-status system that monitors a soldier's vital signs, location, need for water and other critical indicators. Smart wireless sensors and Radio-frequency Identification (RFID) tags are mounting in use and becoming smaller and more powerful. (Halal, 2008). The use of these systems would flourish in the law enforcement world. To have the aptitude to track wanderers and people with disabilities, such as autism and Alzheimer's, would save lives and provide valuable assistance to the law enforcement community.

4.1. Nanotechnology: Locating the Missing

In April 2010, the community of Santa Cruz had been smacked with two unfortunate elderly missing person deaths. Both were senior citizens suffering from Alzheimer's and had walked away from their homes. They were placed in to live safely. The woman(one of them) had been placed in a care facility on a Friday evening. Within the first 30 minutes of her stay, she walked out the front door without detection. The facility was located in a county adjacent to the county for which she had inhabited. She rambled for several days until a resident found her lifeless body in an agricultural ground ten miles north of the facility her family had planned her to live. The second elderly missing man suffered from dementia and like the woman, he rambled out of a care facility and was found deceased several days later. In these two incidents, law enforcement worked tirelessly to locate a missing autistic man in the same month. Erik was a 30-year-old autistic man who walked away from home without a trace. Not long after his disappearance, friends and community volunteers posted fliers all over the county. Sadly, Erik's body washed ashore on a neighboring beach on the last day of the month. Had the families of the missing (and the police charged with finding them) been able to capitalize on emerging RFID and similar nanotech devices, any of the three might have been found alive. John P. Kotter, the author of Leading Change, said: "Vision refers to a picture of the future with some implicit or explicit commentary on why people should strive to create that future." (Kotter, 1996). Nanotech is the epitome of envisioning a future state, and then acting on it. Nanotechnology promises smaller and faster, accurate computers Global Positioning Systems (GPS), batteries with long life and sensors that are incredibly sensitive and affordable (Indian Police Journal, 2007). Enhancing the present RFID technology with faster, more powerful, more sensitive and longer lasting nanotechnology may have future uses that will benefit law enforcement's capability to locate missing people in the quickest manner possible.

4.2. Social Implications and Ethical Considerations

To consider the social and ethical propositions of utilizing nanotech for law enforcement investigations, a group of subject matter experts convened in September 2009. This group identified several concerns regarding the government's use of this incredibly invasive technology. The most pertaining to trends recognized were:

- Ethical usage; law enforcement use of advancing technology must be ethical and transparent. The panel expressed concerns while relating a nanotechnology scenario that could resemble the L.A.P.D. Rampart scandal if there is no transparency about tools and usage
- Civil rights protections: Nanotechnology promises higher quality surveillance systems and concerned was expressed that the technology may be used to obtain information that law enforcement might not otherwise have access to. Privacy rights guaranteed by the Fourth Amendment of the Constitution must be considered and protected
- Information security; concerned was expressed about the information obtained through the use of nanotechnology and that it may be subject to sabotage and theft. Moreover, the panel was concerned that information obtained with nanotech could be sold or used inappropriately by private enterprise and corporations.

- Reliability of the information; will evidence and information obtained with nanotechnology be reliable and who will have the task of qualifying the information.
- Regulatory concerns; what governmental agency will be tasked with regulating the usage of nanotechnology and will that agency have the knowledge base and savvy to do so.

The experts also talked about events that could influence law enforcement's use of nanotech. Chief amongst their concerns were incidents indicating nanotech was unreliable in practice. The adverse impact on privacy rights might prompt legislation & court decisions to limit it, and that exposure to nanotech technologies was determined to be destructive to human or environmental health. The panel did recognize means by which policing and the nanotech industry might alleviate concerns on the near horizon. The recommendations incorporated the development of a law enforcement nanotech research team. Advocating for the establishment of a law based team of nanotechnology experts that will identify the most suitable tools for acquirement and that these tools best meet law enforcement instant needs. As law enforcement agencies obtain new technology, initial and recurring training must be accentuated. Law enforcement agencies will need to account for the protection of nanotechnology tools and create a team of technology experts to support them with this task. Assessing and writing protocols will aid law enforcement leaders with standards and expectations for the use of nano-tools. Newly endorsed laws that regulate governmental use, as well as the creation of a regulatory agency, will provide the public with the translucent use of nanotechnology by governmental entities. Ultimately, providing secure information storage systems to make sure the information obtained through the use of nanotechnology is an ethical and legal fashion and that only those who have legal access to it, can. By scheduling for future law enforcement technological tools and taking the appropriate steps to ensure ethical and reliable usage, the public intellect of security should increase and acceptance should follow. According to Cowper (Policing 2020), in spite of our reluctance, things will continue to change. Tomorrow's policing world will be essentially different than the one we know today. Technology has always generated reflective changes in the way people live, culture, social norms, business processes, organizational structures, and government policies. Policing will turn into more difficult in the next decade and a half, not less, and in order to understand how we will need to operate with the tools available in the future. Therefore, we must try to envision how technology will have changed the world from the way it is today(Policing 2020, 2007). Although there are many promises nanotechnology may offer law enforcement, most importantly law enforcement leaders must take action today to prepare for these powerful crime-fighting tools.

REFERENCES

[1] Russell P. Narcotic Drugs and Psychotropic Substance Act, 1985. *The Indian Journal of Social Work*. 2016; 50(1):119-24.

[2] Gurule J. The 1988 UN Convention against Illicit Traffic in Narcotic Drugs and Psychotropic Substances-A Ten Year Perspective: Is International Cooperation Merely Illusory. *Fordham Int'l LJ*. 1998; 22:74.

[3] P. Ram Manohar. Smoking and Ayurvedic Medicine in India. *Smoke*, 68–75.

[4] *The Opium Act*, 1952.

[5] *The Opium Act*, 1878.

[6] *The Dangerous Drugs Act*, 1930.

[7] *A Detailed Analysis of the National Drugs and Psychotropic Substances* https:// blog.ipleaders.in/ a- detailed- analysis- of- the-national-drugs-and-psychotropic-substances-act/ (last visited, Aug 29, 2018).

[8] Russell P. Narcotic Drugs and Psychotropic Substance Act, 1985. *The Indian Journal of Social Work*. 2016; 50(1):119-24.

[9] Ibid.

[10] Sec. 2 (xv) of the NDPS Act.

[11] Russell P. Narcotic Drugs and Psychotropic Substance Act, 1985. *The Indian Journal of Social Work*. 2016; 50(1):119-24.

[12] Russell P. Narcotic Drugs and Psychotropic Substance Act, 1985. *The Indian Journal of Social Work*. 2016; 50(1):119-24.

[13] *Arif Khan v. State of Uttarakhand*, 2018 (2) RCR (Criminal) 931.

[14] *Union of India v. Mohanlal*, (2012) 7 SCC 719

[15] *Mohinder Singh v. State of Punjab*, (1983) 3 SCC 470.

[16] Russell P. Narcotic Drugs and Psychotropic Substance Act, 1985. *The Indian Journal of Social Work*. 2016; 50(1):119-24.

[17] *Kishan Chand v. State of Haryana*, AIR 2013 SC 357.

[18] *Sukhdev Singh v. State of Haryana*, AIR 2013 SC 953.

[19] Russell P. Narcotic Drugs and Psychotropic Substance Act, 1985. *The Indian Journal of Social Work*, 2016;50(1):11924.5.

[20] *Myla Venkateswarlu v. State of A. P*, AIR 2012 SC 1619.

[21] *Joseph Fernandez v. State of Goa*, AIR 2000 SC 3502.

[22] *Raju Premji v. State of Punjab*, AIR 2009 SC (Supp) 852.

[23] Russell P. Narcotic Drugs and Psychotropic Substance Act, 1985. *The Indian Journal of Social Work*. 2016;50(1):11924.5.

[24] *State of Rajasthan v. Daul alias Daulat Giri*, AIR 2009 SC (Supp) 1592.

[25] Ibid.

[26] Russell P. Narcotic Drugs and Psychotropic Substance Act, 1985. *The Indian Journal of Social Work*. 2016;50(1):11924.5.

[27] *Abdul Rashid v. State of Bihar,* AIR 2001 SC 2422.

[28] Russell P. Narcotic Drugs and Psychotropic Substance Act, 1985. *The Indian Journal of Social Work*. 2016;50(1):11924.5.

[29] *Balbir Kaur v. State of Punjab*, AIR 2009 SC 3036.

[30] *Krishan Mohar Singh Duggal v. State of Goa*, AIR 1999 SC 3842.

[31] *Valsala v. State of Kerala*, AIR 1994 SC 117.

[32] *A Detailed Analysis of the National Drugs and Psychotropic Substances* https:// blog.ipleaders.in/ a- detailed- analysis- of- the-national-drugs-and-psychotropic-substances-act/ (last visited, Aug 29, 2018).

[33] Sharma DB. *Forensic Science in Criminal Investigation & Trials.* Universal Law Publishing Company; 2005.

[34] Id, at 30.

[35] Indian Penal Code, 1860, § 29.

[36] Indian Penal Code, 1860, § 463.

[37] *Ram Chandra v. State of U.P.*, AIR 1957 SC 381.

[38] *Ishwari Prasad Mishra v. Md. Isa,* (AIR 1963 SC 1728).

[39] *Shashi Kumar Banerjee v. Subodh Kumar Banerjee* AIR 1964 SC 529.

[40] *Fakhruddin v. State of M.P.*, AIR 1967 SC 1326.

[41] *Prasad Mahto v. Mt. Jasoda Koer*, AIR 1937 Patna, 328.

[42] *Bhupendra Nath vs Union of India* AIR 1959 HP 29.

[43] *Laxmipat Choraria and Ors v. State Of Maharashtra*, 1968 AIR 938.

[44] *Ratan Lal Soni vs The State of Rajasthan And Ors*, 1994 (1) WLC 679.

[45] *Laxmi Raj Shetty v. State of Tamil Nadu*, 1988 cri lj 1783.

[46] *State of Delhi v. S.J Choudhary*, AIR 1996 SC 1491.

[47] *Ziyauddin Burhanuddin Bukhari vs Brijmohan Ramdass Mehra*, 1975 AIR 1778.

[48] *Cases Reported under Explosives and Explosive Substances Act during 2015*, https://community.data.gov.in/cases-reported-under-explosives-and-explosive-substances-act-during-2015/ (last visited, Aug 30, 2018).

[49] Id, at 42.

[50] *The Explosives Substances Act*, https://lawyerslaw.org/the-explosive-substances-act-1908/ (last visited, Aug 30, 2018).

[51] The Explosives Act, 1908, §2.

[52] *Masha Angami and Anr. v. Government of Manipur* (1967) HC.

[53] *Rajani Kanta Mandal v. State of Bihar*, AIR 1959 Patna 314.

[54] *Chandra Prakash v. State Of Rajasthan*, (2014) 8 SCC 340.

[55] *Giri Raj v. State*, RLW 2005 (4) Raj 2900.

[56] *Public Prosecutor, High Court Of A.P.Hyderabad v. Dorasala Patnam Nadipi Veera Reddy.*

[57] *Mohammad Usman Mohammad Hussain Maniyar And Others vs. State Of Maharashtra*, 1981 AIR 1062.

[58] *Nema Adak v. State of West Bengal*, 1965(1) Cri LJ 160(Cal).

[59] *Balachandra v. State of Maharashtra*, 1968 Cri LJ 1501(SC).

In: Introduction of Forensic Nanotechnology ... ISBN: 978-1-53616-040-6
Editors: R. K. Shukla and A. Pandya © 2019 Nova Science Publishers, Inc.

Chapter 10

NANO-TOOLS FOR ILLICIT DRUG SENSING: ADVANCES AND CHALLENGES IN FORENSIC INVESTIGATION

***Niha Ansari*[1], *Pinkesh Sutariya*[2] *and Alok Pandya*[3,*]**
[1]Institute of Forensic Science,
Gujarat Forensic Science University, Gujarat, India
[2]Department of Chemistry, Bhavan's Shree I.L. Pandya Arts-Science,
and Smt. J.M. Shah Commerce College, Dakor,
Sardar Patel University, Gujarat, India
[3]Department of Physical Sciences, Institute of Advanced Research,
Gujarat, India

ABSTRACT

Illicit drug analysis is an important branch of modern chemistry with many legal and socially relevant implications in forensic investigations. It requires an important role of nanotechnology due to issues on cost-effectiveness, instrumentation constraints and sensitivity of traditional approach. So, driven need to detect the illicit drug, various nano-enabled tools (nanotools) can play a significant role in forensic analysis considering their well-known advantages of sensitivity, selectivity, easy functioning, affordability and capability of miniaturization and automation. Although, the use of nano-enabled screening in forensic practice is still very limited. Thus, this chapter will give an overview of the potential of this new-fangled area of research, which aimed to describe new techniques for "*on-site*" analysis of abused drugs at trace level. It will transform as an future armor in the fields of crime scene investigation.

Keywords: illicit drug, nano-enabled tools, Forensic

1. INTRODUCTION

Drug-facilitated crime (DFC) is significantly enhanced in recent years that includes sexual assault, deliberate maltreatment of the elderly or children, robbery and money extortion under the manipulation of psychotropic substances. DFC are criminal acts carried out by means of administering a substance to a person with the intention of impairing behavior, perceptions or decision-making capacity. Moreover, it also extends to taking advantage of a weakened person, without their consent, after their voluntary intake of an incapacitating substance. Drugs to facilitate crime has been highlighted by a significant increase in reports of DFC worldwide[1-6]. The use, abuse or dependence of addictive substances has been well acknowledged. Figure 1 shows the widely abused drugs globally. According to world drug report by United Nations Office on Drugs and Crime (UNODC), it is estimated that 183,000 drug-related deaths were reported in 2012 [7]. In 2012, it is estimated that some 243 million people were in the age of 15-64 had used an illicit drug mainly a substance belonging to the cannabis, opioid, cocaine or amphetamine. It has seen that during the year 2009-2012, the number of drug users increases significantly.

This report generates worries for our society and creates the opportunity to detect illegal narcotics. It is not only important but challenging for many forensic scientists because of its lack of availability in pristine form in the sample, which is sent for examination. Another challenge for scientists is to detect the drug in trace amounts of biological specimens. A forensic scientist receives different types of biological samples, including blood, saliva, semen, and urine, from the crime scene. These drug tests are generally carried out in two ways i.e., screening and confirmation. Once drugs are detected through screening by spot test kits like immunoassays, Marquis test, etc., samples are then collected and sent to laboratories for confirmation tests. Confirmation requires high sensitivity and selectivity toward drugs, as well as their metabolites, and is frequently carried out by chromatographic and spectroscopic techniques [8]. Chromatographic techniques like Gas Chromatography/Mass Spectrometry (GC/MS), Liquid Chromatography/ MS(LC/MS), High Performance Thin Layer Chromatography (HPTLC), High Performance Liquid Chromatography (HPLC), Reverse Phase/HPLC, HPLC/MS/MS etc. and spectroscopic techniques like Fourier transform infrared spectroscopy(FT-IR), Nuclear magnetic resonance (NMR), Mass Spectrometry (MS), UV-visible and Raman spectroscopy have been found in forensic laboratories as complementary techniques[8]. However, these techniques are limited to the laboratories and also suffer from a number of limitations in terms of selectivity, sensitivity, and stability. Moreover, they are tedious, costly and time-consuming. Nowadays, these drawbacks are overcome by the developing of sensors. such as the development of affinity-based biosensors and their usefulness in the field of forensic analysis. DNA based sensors also have irrupted in the field of forensic toxicological analysis showing their high usefulness such as detection of pathogen micro-organisms. Furthermore, aptamer-based sensor has been also developed for different forensic applications[9]. In the case of enzyme biosensors, diversity of methods developed for species of interest in forensic chemical analysis, In few cases, these methodologies have been applied to the complex biological samples which are currently analyzed in the forensic laboratories. However, several enzymes that can be employed to design biosensors with a demonstrated usefulness for the detection of inorganic poisons such as arsenic or cyanide, and of chemical or biological war agents, among others.

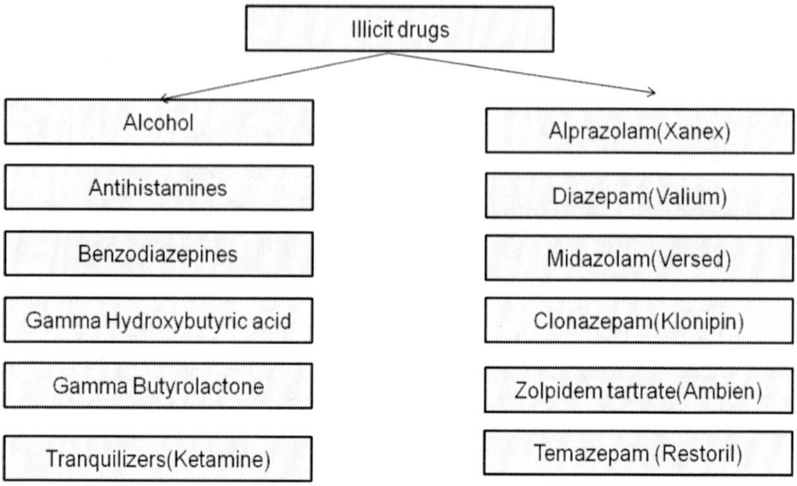

Figure 1. Illicit drugs in the world.

It is a well-known fact that desired nanostructured material can be designed and used for making nanosensor which offers excellent signal amplifier for chemical and biological sensing.It is due to their unique optical, electrical, and thermal properties as well as catalytic properties. Nanosensors usually show enhanced biocompatibility, sensitivity, and specificity over most conventional sensingsystems [10]. However, the use of nanosensors in forensic practice is still very limited, as it is reflected by the scarce number of articles on this topic reported on biosensors. In this paper, we review the nano-intervention for forensic relevant analysis, in particular, those most prominent drugs used in a facilitated crime. Any type of detection strategy can be considered, although the special emphasis is made in optical and electrochemical transduction based methods. Therefore, the chapter has been organized according to the main areas of forensic analysis involving forensic toxicological analysis and theapplications related to drug screening. Despite their proven utility, there is a continuous demand for the improvement and optimization of current analytical techniques to detect and identify existing and emerging illegal substances with enhanced sensitivity and selectivity[11]. Therefore, this chapter aimed to describes the wat that nano-intervention can to improve the sensitivity and selectivity towards the analysis of illicit drugs.

Nanotechnology is creating a wealth of new materials at the nano level (~1 to 100 nm)and manufacturing possibilities, which in turn will profoundly impact our economy, our environment, and our society. Research on nanomaterials and nanodevices are becoming more popular because of their electrical, mechanical and chemical properties. Nanodevices such as Nanobiosensor, nanochip, nanoprobes, etc. have been developed for various applications in the life science field for hybridization assays [12-14], interaction studies [15-18], disease diagnosis [19-22], metal detection[23-25], etc. and showed excellent sensitivity and selectivity. So, these nano-enabled devices can be made for forensic analysis which significantly improves sensitivity and selectivity. Nanotechnology also allows the miniaturization and the integration of biomolecules, transduction systems, electronics, and microfluidics incomplexnanobiosensorarchitectures.This architecture allows high throughput analyses as lab-on-chip devices for the rapid and low-cost screening of forensic drugs and their metabolites, using small samples of drug-exposed material such as blood, serum, saliva, and plasma [26].

In the construction of nanosensor development, *molecular receptor* plays a key role as a recognition site in construction. More importantly, it also enables highly specific assays for drug screening. Hence, different types of molecular receptors specific to the drug will be addressed in the following sections.

2. Molecular Receptors

Functionalized nanomaterials with molecular receptors are a well known fact in nanoscience.Molecular receptors are that component of a nanosensor which serves as a template for the material to be detected. There can be several materials which can be used as receptors. For drug screening, the biomolecules for CNS (Central Nervous System) depressant drugs are represented in Table 1[27]. This binding sites or recognition sites can play a pivotal role in the construction of nanosensor. *Gamma-aminobutyric acid* (GABA) receptor can be interesting biomolecule for detection of the barbiturates, benzodiazepines, ethanol, and picrotoxins as it binds with GABA receptor. Similarly Phencyclidine, a CNS depressant drug is widely abused and it blocks the action of NMDA receptor. So, the N-methyl-D-aspartate (NMDA) receptor can be great recognition site for detection of phencyclidine. Similarly, narcotics are the widely abused drugs among all other drugs due to their hallucinogenic action. Fast and accurate detection of such drugs from biological fluids will be useful in emergency cases. Similarly, some binding sites for narcotic drugs are described in Table 2 [28]. Mostly, morphine, codeine, heroin, and cannabis are widely abused drugs. Many death cases have been reported due to an overdose of anti-depressant drugs as well. Tricyclic and mono-amino oxidase inhibitors (Table 1) are the class of anti-depressant drugs which inhibits the enzyme named Phosphodiesterase and Monoamine-oxidase respectively. So, determining the conversion of this enzyme could be a great innovation for the detection of these drugs. Serotonin and Lysergic acid derivatives bind with the 5-HT receptor, so this receptor can be a useful recognition site for detection of these drugs. Nicotine and cocaine are the most widely abused CNS stimulant. Nanosensor can be constructed for their detection based on the biorecognition site. Nicotine act by Nicotinic Acetylcholine receptor and exert its pharmacological action, thus this biomolecule can be the good binding site for nicotine. The other CNS stimulant drugs and heir recognition site has given in Table 3 [29].

Table 1. Recognition site for detection of CNS depressant drugs

Target Drug	**Recognition site**
Barbiturates	$GABA_A$ receptor
Benzodiazapines	$GABA_A$ receptor
Ethanol	$GABA_A$ (modulator)
Phencyclidine	NMDA (antagonist)
Tricyclic anti-depressants	Phosphodiesterase inhibitor
Mono-amino oxidase inhibitors	Monoamino-oxidase
Serotonin re-uptake inhibitors	5-HT receptor activation
Lysergic acid and its derivatives	Bind with serotonin in nerve ending, 5-HT2 (partial agonist)

Table 2. Recognition site for detection of narcotic substances

Target Drug	Recognition site
Morphine, codeine	μ1 and μ2 receptor
Methadones	NMDA receptor
Phenylpiperadines (ex-fentanyl, alfentanyl)	μ receptor
Heroin	μ1 and μ2 receptor
Levorphanol	μ receptor
Cannabis (Marihuana)	CB1, CB2 agonist
THC	CB1-receptors

Table 3. Recognition site for CNS stimulant drug detection

Target Drug	Recognition site
Amphetamines	Effect on α and β receptor
Atropine	M2 (muscarinic) antagonist
Cocaine	Dopamine transporter
Xanthines	Phosphodiesterase inhibitor
Premoline	Dopamine receptor antagonist
Methylphenidate	Dopamine receptor antagonist
Picrotoxin	GABAA receptor
Doxapram	Carotid chemoreceptor inhibitor
Strychnine	Binds to glycine receptors, also inbibitor of Ach receptor
Nicotine	Nicotinic Ach

3. Nanotoolsfor the Detection of Illicit Drugs

3.1. Codeine Detection

Codeine (3-methylmorphine) is an alkaloid separated from Opium. Codeine is a short-acting narcotic often used for the treatment of pain relief. It belongs to the class opiates. Codeine causes the release of neurotransmitters that stimulate the reward center of the brain and provides an overall sense of calm and feelings of pleasure. This kind of pleasure can lead to both psychological and physical dependence. Codeine is highly addictive. This drug is generally available as a single agent or in combinations with other analgesics drugs such as acetylsalicylic acid, ibuprofen, and caffeine in pharmaceutical tablets. When used in conjunction with alcohol, codeine can enhance their respective effects and increase the potential for a drug overdose. Alcohol inhibits the metabolism of codeine, thereby increasing codeine's effects.

In recent years, several methods for the detection of codeine using nanotechnology have been developed. XiaofangNiu et al. [30] reported a novel aptamer biosensor for the ultrasensitive codeine detection based on gold nanoparticles (AuNPs)/polyamidoaminedendrimer-modified screen-printed carbon electrode. They used Polyamidoaminedendrimers (PAMAM) to absorb gold nanoparticles (AuNPs) and modify the surface of screen-printed carbon electrodes (SPCEs). The codeine aptamer was immobilized on the PAMAM/glutaraldehyde (GA)/chitosan (CHIT)-modified electrode through Au–SH affinity. Codeine obstructed the electron transfer of electrochemical probe $K_3Fe(CN)_6/K_4Fe(CN)_6$, hence can be used to detect codeine. The detection limit was found to be 3×10^{-13}mol L^{-1}. The method was also used to detect codeine directly in blood serum and other aqueous solutions. Recently, ForoozanHasanpour et al. prepared Copper ferrite-copper oxide ($CuO/CuFe_2O_4$) nanoparticles as a semiconductor composite with p-n junction by co-precipitation reaction. The synthesized nanocomposite then used as a modifier in carbon paste electrode (CPE) for the sensitive quantification of acetaminophen and codeine in the biological fluids. The electrode showed high reproducibility and sensitivity for routine analysis of acetaminophen and codeine in urine and plasma samples. The Limit of detection for codeine was found to be 0.01 mmol L^{-1}. Furthermore, no interference was noticed.

Ali Babaei et al. [31] reported the electrochemical oxidation of codeine by the application of a novel glassy carbon electrode modified with hydroxyapatite-Fe_3O_4 nanoparticles/multiwalled carbon nanotubes composite (HA-FeNPs-MWCNTs/GCE) [Figure 2] The results showed that combination of HA-FeNPs and MWCNTs as a composite in the modified electrode at optimum conditions provides very high sensitivity for the simultaneous determination of codeine due to excellent electrocatalytic activity and high surface area of the composite. At a pH of 7.5, there was an increase in selectivity and sensitivity of the proposed method. The interfering study was done along with the other species namely ascorbic acid, uric acid, l-Glutamic acid, L-Alanine, Aspartic acid, l-Histidine, Dopamine, Tryptophan, Diclofenac, and the results showed no significant interference with the determination of codeine. The differential pulse voltammetry results showed that the obtained anodic peak currents were linearly proportional to the concentration in the range of 0.1–28 μM and with a detection limit of 22 nM for codeine. Application of the proposed sensor for the determination of codeine in real samples showed satisfactory results, without the requirement of sample pre-treatment or time-consuming extractions. The simple fabrication procedure, high sensitivity, excellent selectivity, good reproducibility, and high speed suggested that the proposed sensor can be utilized for practical applications.

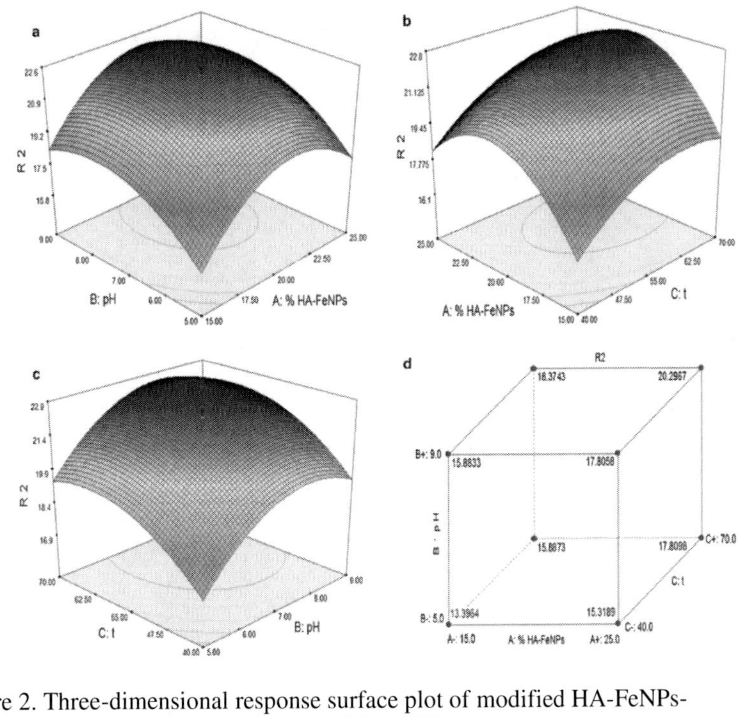

Figure 2. Three-dimensional response surface plot of modified HA-FeNPs-MWCNTs/GCE for the detection of Codeine [31].

The recent development of nanomaterials and nanotechnology has opened new horizons in designing integrated bio/electrochemical systems. The nanoscale design of the electrode surfaces provides novel and sometimes unique properties due to the ability of the intrinsic properties of the nanomaterials employed, the ability to control the architecture of the electrode interface at the nanoscale, or both. The nanostructured electrodes, or porous materials held at the electrode surfaces enhance the surface area/volume ratio by several orders of magnitude. This results in improved interactions with external reagents and/or the immobilization of different functional moieties.

A nanocomposite of graphene (Gr) and Cobalt ferrite ($CoFe_2O_4$) nanoparticles weresynthesized by Abbas Afkhami et al. [32] in order to fabricate a modified carbon paste electrode. The unique behavior of the $CoFe_2O_4$/Gr/CPE allowed electrochemical determination of codeine. The proposed method showed no interference effects of glucose, ascorbic acid, caffeine, naproxen, alanine, phenylalanine, glycine, and others. The electrode showed a fast electron transfer reaction and without electrode fouling problems. High stability, selectivity, and reproducibility, as well as low detection limits, ease of preparation, low cost, and surface renewal, made the proposed nanocomposite electrode ideal for the detection Codeine in real samples. The low detection limits werefound to be 0.011 μM based on three times of the standard deviation of the blankover sensitivity (3 s/m).

Based on incorporating $ZnCrFeO_4$ magnetic nanoparticles into multi-walled carbon nanotubes paste matrix (MWCNTs/ZnCrFeO4/CPE), a chemically modified electrode was prepared for the determination of codeine by M. Tai et al. [33]. The detection limits (S/N =3) was found to be 0.01 μmol L^{-1}. The method was also successfully employed as a selective, simple, and precise method to determinate AA, AC, and CO in pharmaceutical and biological samples. The edge plane-like sites/defects of MWCNTs along with the magnetic effect of $ZnCrFeO_4$ are responsible for amplifying the oxidation signal of CO. $ZnCrFeO_4$ nanoparticles modified carbon paste electrode also showed maximum sensitivity towards determination of CO compared to spinel ferrite nanoparticles such as $MgFe2O_4$, Fe_3O_4, $NiFe_2O_4$, $ZnFe_2O_4$, and $CoFe_2O_4$. The presence of ZnCrFeO4 offers advantages in terms of analytical figures of merit such as enhanced sensitivity, low limit of detection and wider dynamic range than the non-MNPs modified electrode. Similarly, Mohammad Hossein Mashhadizadeh and Fatemeh Rasouli [34] fabricated electrochemical sensor via TiO_2 nanoparticles onto a carbon paste electrode. The Cyclic voltammetry and differential pulse voltammetry analysis was done of the modified electrode toward codeine. The effects of pH, modifier amount, pulse amplitude, and a scan rate of potential were also examined. The detection limits of 0.018 $mmolL^{-1}$ were achieved. The proposed voltammetric sensor was successfully applied to the determination of codeine in human plasma serum samples.

Ensafi AA et al. [35] synthesized porous silicon/palladium nanostructure and used as a new electrode material for the codeine detection.They assembled Palladium nanoparticles were on porous silicon (PSi) microparticles by a simple redox reaction between the Pd precursor and PSiin an aqueous solution of hydrofluoric acid. A linear range of 1.0 – 700.0 μmol L–1 was achieved with detection limits of 0.3 μmol L^{-1}. The proposed method was used for the determination of COD in blood serum, urine and pharmaceutical compounds.

3.2. Morphine Detection

Morphine (MO) is a phenolic compound and an alkaloid which is highly effective and preferred as a drug to relieve severe pain in patients, especially those undergoing a surgical procedure which can cause disruption in the central nervous system. Morphine is an active principle ingredient purified from a plant source and has at least 50 alkaloids of several different types present in opium, Poppy Straw and other poppy derivatives. Morphine can lead to addiction; tolerance and psychological dependence rapidly. To prevent overdose-induced toxicity, monitoring of the therapeutic levels of MO concentrations in blood or urine in patients is a very important in clinical medicine and epidemiological investigation for drug abuse control as well as in forensic cases. The most common analytical techniques currently used to determine MO concentrations are high-performance liquid chromatography (HPLC), GC–mass spectroscopy, liquid chromatography with UV detection, laser-induced fluorescence detection, LC-mass spectrometry detection, capillary electrophoresis with mass spectrometry and surface plasmon resonance based immunosensors [36]. Morphine is an important analgesic, narcotic and can be used as an indicator of drug abuse. Thus, there is a need to develop alternative, simple, and easy approaches to monitor these opiate drugs in biological samples. Various authors have reported the use of nanotechnology for the detection of MO from various biological fluids.

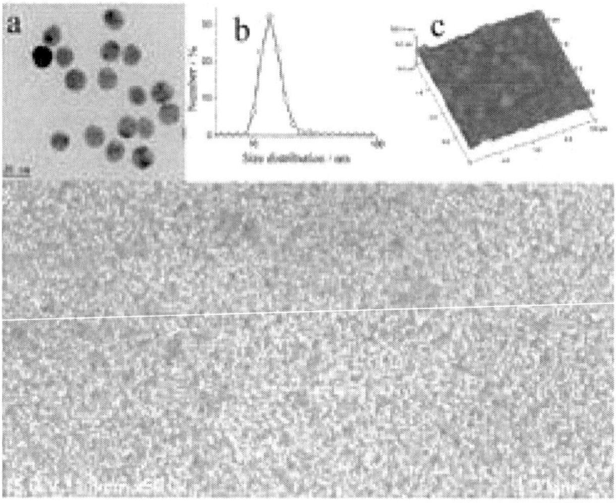

Figure 3. Structural analysis of gold nanoparticles used for the detection of Morphine [37].

Yang Ya et al. [37]. has reported an electrochemiluminescence (ECL)-based morphine (MO)Immunosensor. They have developed label-free morphine immunosensor based on direct adsorption of antibody onto the surface of gold nanoparticles functionalized ITO glass [Figure 3]. They had attached gold nanoparticles onto the surface of indium tin oxide coated glass which acted as a matrix to directly immobilize the MO-antibody as sensing host via the adsorption. With a totally label-free mode, the ECL intensity of luminol present on the sensor decreased along with the concentration of target MO solution after the immuno-incubation. The linear range was found to be 2 to 200 ng/mL and a detection limit of 0.82 ng/mL (S/N=3). It has been used to zanalyzethe MO in spiked urine samples with satisfactory recoveries to check its applicability. This label-free sensor has the potential possibility as a disposable, portable and mass-producible device [37].

Ali Babaei et al. developed a glassy carbon electrode modified with carbon nanotube/chitosan (MWCNTs-CHT/GCE) for the sensitive voltammetric determination of MO. The electrochemical response was investigated by cyclic voltammetry (CV), differential pulse voltammetry (DPV) and chronoamperometry (CA). Under optimum conditions, the sensor provides a linear DPV response in the range of 2.0×10−6 to 1.0×10−4 M with a detection limit of 2.4×10−7 M for MO. The developed composite film combines the advantages of ease of fabrication, an efficient electron transfer reaction together with well-defined oxidation peaks for determination of MO. Moreover, the sensor has wide linear range, speed, low cost, sub-micromolar detection limit and simple approach for redevelopment of the electrode surface [36].

Figure 4. Structural analysis of gold nanoparticles used for the detection of Morphine [39].

Atta's group developed an AuNPs and Nafion modified CPE (Carbon paste electrode) for voltammetric determination of MO. The modified electrode proved to be efficient for the electrocatalytic oxidation of morphine. The peak current of morphine was improved by AuNPs due to the higher electroactive surface area, and by increasing the electron transfer kinetics (Figure 4). The limit of Detection was 1.33 nM, however, the LOD of morphine in the urine sample was 87.2 nM, which indicates that this method is not selective owing to the interferences present in urine matrix [38]. In 2012, again Atta et al. used gold nanoparticles/ferrocene modified CP-electrode (GNFMCPE) for the MO detection. The electrochemistry of MO was investigated by cyclic voltammetry (CV), differential pulse voltammetry (DPV) and electrochemical impedance measurements (EIS). The morphine concentration was measured in the linear range of 1.0×10^{-6} to 18.0×10^{-4}mol L^{-1} with a detection limit of 3.507 x10-9 mol L^{-1} and a correlation coefficient of 0.9994. Here they demonstrated the selective determination of MO in the presence of AA and UA in 0.04 M B-R buffer (pH 7.4) using GNFMCPE. They found that the bare CP-electrode failed to resolve the voltammetric signals of a mixture of MO from ascorbic acid (AA) and uric acid (UA). While, the GNFMCPE enhanced the current signals of MO, the oxidation peak current of AA was disappeared and the low response of UA obtained [39]. In 2014, Atta's group again reported the use of the electrochemical sensor for MO detection in presence of sodium dodecyl sulfate (SDS) by utilizing gold nanoparticles over poly(3,4-ethylene-dioxythiophene) modified gold electrode (Au/PEDOT-Au_{nano}...SDS). They found enhanced electrocatalytic activity towards MO oxidation at Au/PEDOT-Au_{nano}...SDS compared to Au/PEDOT and Au/PEDOT-Au_{nano}. The MO concentration could be measured in the linear range of 2 μM to 18 μM with a correlation coefficient of 0.9983 and very low detection limit of 0.428 nM. Furthermore, they also showed the simultaneous and selective determinations of MO, ascorbic acid and uric acid, MO and dopamine and MO and atropine with high sensitivity and reproducibility [40].

Gold nanoparticle (AuNP)-based optical assays were reported by Sonu Gandhi et al. for the detection of MO. They help us to examine the molecular phenomenon easily with a change in the color of AuNPs (Figure 5). They developed a dipstick using an AuNP-labeled single-chain fragment variable (scFv) antibody for the detection of morphine using phage display-based antibody library. Herein, they connected immunoglobulin variable regions of heavy (VH)- and light (VL)-chain genes via a glycine-serine linker isolated from the murine immune repertoire and cloned into the expression vector pIT2. The scFv was produced in Escherichia coli HB2151, yielding a functional protein having a molecular weight of approximately 32 kDa. The morphine scFv was labeled with gold nanoparticles and hence used as an optical immunoprobe in a dipstick. The developed recombinant antibodies bind very specifically to morphine and monoacetyl morphine. ELISA-based results indicated that morphine-specific scFv antibodies possess the potential to be used as a molecular affinity interface receptor and can be used as a biosensor device. The detection range was found to be 1–1000 ng mL^{-1} with a limit of detection (LOD) of 5 ng mL-1 under optimal conditions, and the IC50 value was 14 ng mL-1 for morphine. An optical, fast, and field-applicable dipstick immunosensor kit was developed for morphine which was capable of specifically binding to free morphine and its analogs in a solution in less than 5 min and could be useful for on-site screening of a real sample in blood, urine, and saliva [41].

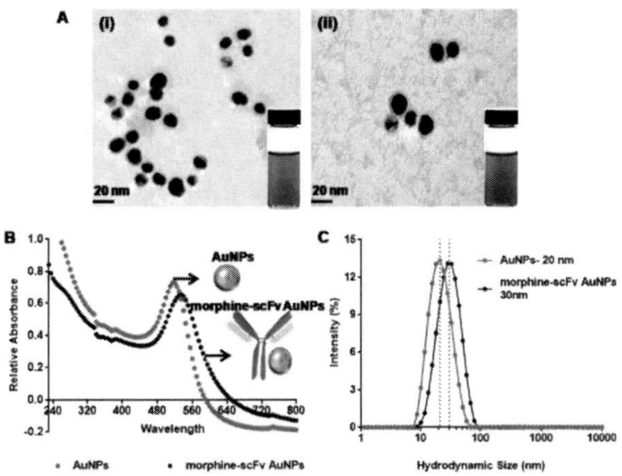

Figure 5. Detection of Morphine by morphine-scFvAuNPs [41].

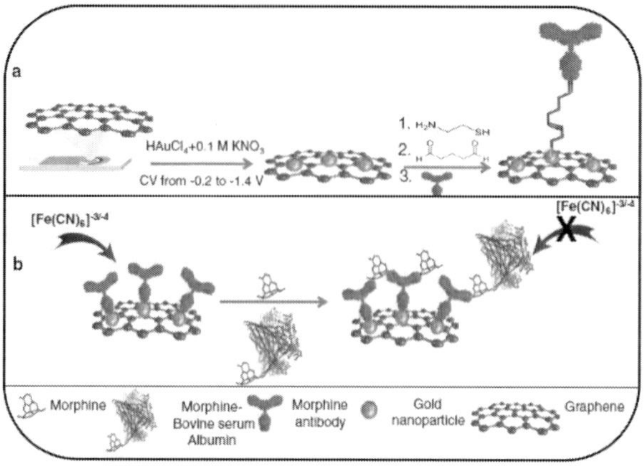

Figure 6. Voltammetricimmunosensor for MO developed on a graphene screen screen-printed electrode (GSPE) modified with gold nanoparticles (AuNP) [42].

A voltammetricimmunosensor for MO was developed on a graphene screen-printed electrode (GSPE) modified with gold nanoparticles (AuNP) by S. Eissa (Figure 6). Further cysteamine was self-assembled on the gold-modified electrodes via thiol interaction to introduce terminal amino groups to the electrode surface. The electrodes were used to fabricate the immunosensor by covalent immobilization of antibodies against MO. The sensor basically worked on the basis of a competition between MO and the morphine-bovine serum albumin conjugate for the immobilization of antibodies on the sensor surface and thus resulting change in the square wave voltammetry reduction current using the hexacyanoferrate system as an electrochemical probe. The competitive assay enabled the detection of MO with only single incubation and washing steps. The AuNP modified graphene electrodes lead to enhancement of the electrode surface area and electron transfer efficiency. The competitive immunoassay enables sensitive and selective detection of MO in the range of 0.1 to 100 ng·mL−1, with a 90 pg·mL−1 detection limit [42].

Sara Dehdashtian et al. [43] and HasanBagheri et al. [44] also showed the used of chitosan-coated magnetic nanoparticle (CMNP) and Zn_2SnO_4-graphene nanocomposite for detection of MO. They modified carbon paste electrode with the chitosan-coated magnetic nanoparticle (CMNP) and Zn_2SnO_4-graphene nanocomposite, respectively. Both have utilized cyclic voltammetry,Chronoamperometry and differential pulse voltammetry as diagnostic techniques to study the electrooxidation of MO on modified carbon paste electrode. The detection limit for MO was found to be 3 nM in case of CMNP and was 0.011 μmol L-1 in case of Zn_2SnO_4-graphene nanocomposite.

HasanBagheri et al. also did interference study and found no significant interference of Ca^{2+}, glucose, lactose, sucrose, ascorbic acid, acetaminophen, ethanol, noscapine and others in the detection of analytes. In the year 2014, WenjuanFei et al. showed the use of gold electrode which was modified with a nanocomposite film containing self-assembled polyamidoamine (PAMAM) CdSquantum dots and electrodeposited gold nanoparticles (Au-NPs) (Figure 7). Due to the synergistic effect of the modified quantum dots and the electrodeposited Au-NPs, the electro-chemiluminescent (ECL) response was intensely enhanced. The immunoreaction between morphine and anti-morphine antibody showed a decrease of the ECL signal because of steric hindrance, under optimal experimental conditions. The linear range was found to be 0.2 to 180 ng•mL−1, with a detection limit as low as 67 pg•mL−1. The sensor was successfully applied for the determination of morphine in blood plasma [45].

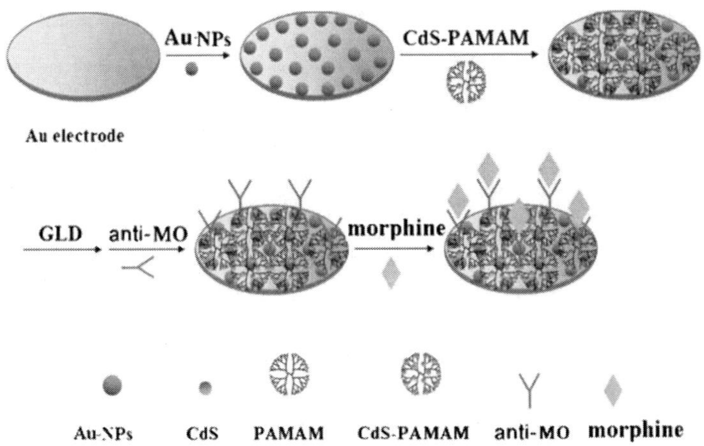

Figure 7. Schematic representation of the detection of morphine by Au electrode by using CdS-PAMAM [45].

Ali A. Ensafi et al. also showed the use of a carbon paste electrode modified with multiwall carbon nanotubes (MWNTs). Further, they utilized ILs (n-hexyl-3-methylimidazolium hexafluorophosphate) to fabricate a carbon nanotubes paste electrode for the detection of morphine. Cyclic voltammetry, chronoamperometry, and electrochemical impedance spectroscopy (EIS) were used to study the electrochemistry of morphine at the surface of the modified electrode. The cyclic voltammogram showed an irreversible oxidation peak at 0.48 V (vs. Ag/AgClsat), which corresponded to the oxidation of MO. The detection limit of 0.02 $\mu mol\ L^{-1}$ was found using differential pulse voltammetry. The proposed sensor was successfully applied for MO determination in biological and pharmaceutical samples such as human urine and injection solution [46].

M. Masteri-Farahani et al. [47] developed an optical nanosensor for measurement and detection of MO using CdS quantum dots (CdS-QDs) functionalized USING antimorphine antibody.The prepared nanosensor (CdS-QDs-antimorphine) was based on the selective host-guest interaction between morphine and antimorphine antibody. The modification of CdS-QDs with antimorphine antibody quenches the primary fluorescence of CdS-QDs according to the Forster resonance energy transfer (FRET) mechanism and the next interaction with morphine turns on the fluorescence of CdS-QDs-antimorphine. FRET process avoids the recombination of the electron-hole pairs on the primary photo-excited CdS-QDs-antimorphine and leads to the radiationless fluorescence quenching of the CdS-QDs antimorphine. The dependence of fluorescence enhancement on the MO concentration can be used for the measurement of morphine in the real samples.

3.3. Cocaine Detection

Addiction to cocaine is a serious universal problem. Cocaine functions is a strong stimulant which acts on the central nervous system and misuse of cocaine can cause the severe side effects on human, such as organ damage, anxiety, cardiac arrest and spread of human immune deficiency.

Different techniques have been applied for detection of cocaine including enzyme-linked immune sorbent assay (ELISA), high-performance liquid chromatography (HPLC), gas chromatography-mass spectrometry (GC-MS) and radioimmunoassay. These analytical approaches are time-consuming, require sample pre-treatment, sophisticated instruments expensive and require trained personnel. Simple and sensitive methods for detection of cocaine are essential for both law enforcement and clinical diagnosis.

Mahmoud Roushani and FaezehShahdost-fard constructed an aptasensor by covalent immobilization of aptamer-functionalized AgNPs on the multiwalled carbon nanotubes/ionic liquid/chitosan (MWCNTs/ IL/Chit) nanocomposite as the sensor for cocaine detection.They used riboflavin (RF) as a redox probe in the development of the electrochemical aptasensor. The acceleration of the electron transfer kinetics related to the reduction of RF was achieved by AgNPs. The presence of nanocomposite resulted in prominent biocompatibility, good electrical conductivity, and vast surface area functions as a signal amplifier to facilitate the immobilization of the 5'-NH_2-3'-AgNPs terminated aptamer on the surface of the electrode. The proposed method showed high sensitivity and selectivity with LOD of 150 pM and was used for diagnosis of cocaine in serum samples [48].Seyed Mohammad Taghdisi et al. developed an electrochemical aptasensor for sensitive and selective detection of cocaine, based on single-walled carbon nanotubes (SWNTs), gold electrode and a complimentary strand of aptamer (CS).The developed aptasensor possess the properties such as large surface area, high affinity, and selectivity toward its target, high electrochemical conductivity and stronger interaction of SWNTs with single-stranded DNA (ssDNA). In the presence of cocaine, a strong electrochemical signal was achieved as the aptamer binds to cocaine and leaves the surface of electrode thus a large number of SWNTs bind to CS-modified electrode. The limit of detection (LOD) was 105 pM and in serum, LOD was as low as 136 pM [49].

Ahmad SarreshtehdarEmrani et al. [50] in 2016, developed a novel fluorescent aptasensor for detection of cocaine based on the combination of the hairpin structure of a complementary strand of aptamer (CS), silica nanoparticles coated with streptavidin and gold nanoparticles. The use of two kinds of nanoparticles with opposite effects on fluorophore (FAM) and the hairpin structure of CS resulted in a very significant difference in the fluorescence intensity of the designed aptasensor in the presence and absence of cocaine. A 44 per ssDNA was utilized as targeting ligand which binds to cocaine with high affinity. In the absence of cocaine weak fluorescence emission was noticed as FAM was in close proximity to the surface of AuNPs and in the presence of target very strong fluorescence emission was noticed as FAM comes to close proximity to the surface of SNPs because of the formation of the hairpin structure of CS. The limit of detection (LOD) was as low as 209 pM. Moreover, the designed aptasensor was also utilized to detect cocaine in serum with a LOD as low as 293 pM.

Kun Li et al. [51] reported a novel nanoplasmonic approach for preparation of high-resolution dark-field microscopy (DFM) images of the latent fingerprint (LFP) for detection of cocaine present in LFPs.The cocaine aptamer was divided into two flexible ssDNA fragments which were attached to different AuNPs via Au-S bound. In the presence of cocaine, it acted as linker molecule and reassemble the two fragment of ssDNA into intact aptamer conformation, thus resulting in the aggregation of the AuNPs probes. This aggregation resulted in a colour change from red-to-blue. For detection of cocaine in LFPs, fingers contaminated with cocaine and LFPs were collected and then incubated with aptamer-bound AuNPs and finally observed under DFM. The increase of cocaine concentration from 150 ng to 30 μg, high quantity of AuNPs showed aggregated.

P. Hashemi et al. developed a novel label-free aptasensor based on screen-printed carbon electrode (SPCE). The aptasensor was modified with three-dimensional magnetic reduced grapheneoxide(3D-MRGO)/polyaniline(PA)/gold nanoparticle(AuNP) nanocomposite for impedimetric determination of cocaine. Further, a specific thiolated cocaine aptamer was immobilized onto the surface of the synthesized nanocomposite. The signaling mechanism was based on the increase in the $[Fe(CN)_6]^{3-/4-}$ charge transfer resistance (RCT) as an electrochemical probe in the presence of cocaine. Also to collect 3D-MRGO/PA/AuNP/aptamer on the surface of the working electrode, a new electrochemical cell was fabricated. Under optimized condition, the linear concentration range was found to be 0.09 to 85 nM with a detection limit of 0.029 nM by electrochemical impedance spectroscopy [52].

Zhonghua Chen et al. in their work showed the use of chemiluminescent detection technique coupled with mesoporous silica-based controlled released system for cocaine detection. The MSN support was firstly loaded with glucose, then the positively charged MSN was interacted with negatively charged oligonucleotides (the aptamer cocaine) to close the mesopores of MSN. The cocaine gets binds with its aptamer with high affinity thus the flexible linear aptamer structured will become stems structured and causes the releasing of entrapped glucose into the solution. Further, with the assistant of glucose oxidase (GOx), the released glucose will react with the dissolved oxygen to produce gluconic acid and H_2O_2. The released H_2O_2 will further enhance the Chemiluminescence of luminol in the NaOH solution. The enhanced Chemiluminescence intensity will occur depending on cocaine concentration. The detection limit was found to be 1.43 μM. The proposed method was also successfully applied to detect cocaine in serum samples with high selectivity [53].

3.4. Methamphetamine Detection

Methamphetamine (METH) is the second most widely abused drug in the world.The abuse of such type of illicit drugs may cause severesocietal consequences, such as loss of lives, the health of abusers and higher incidence of crimes. METH is a potent stimulant to the central nervous system and is used to treat obesity and alcoholism. Many instrumental methods are used for its detection including gas chromatography, HPLC, gas chromatography-mass spectrometry (GC-MS), chromatography-mass spectrometry [54]. Kang Mao et al. designed a colorimetric biosensor based on non-aggregation Au@Ag core-shell nanoparticles to detect METH.The biosensor consisted of a specific single strandedDNA (ssDNA) sequence as reporter probe (RP) coated on Au@Ag nanoparticles, a capture probe (CP) conjugated with magnetic beads, and an illicit drug-binding DNA aptamer (Apt). The DNA aptamer could hybridize with both RP and CP, thus creating anAu@Ag-DNA magnetic beads (MBs) sandwich structure. When an external magnetic field was applied, Au@Ag-DNA-MB get removed from the solution, leading to decreased SPR intensity. In the presence of METH, Au@Ag-DNA-MB sandwich structure was not formed due to the higher affinity of the aptamer to the illicit drug than that to complementary ssDNA. The SPR intensity was in correlation with the METH concentrations. The developed method was able to detect as low as 0.1 nM (14.9 ng L^{-1}) METH with negligible interference from other common illicit drugs [55].

Melina Yarbakht& Maryam Nikkhah designed a sensitive bioassay strategy using gold nanoparticles (GNPs) and the aptamers that possess high affinity toward methamphetamine.The aptamer was able to adopt different tertiary structures in the presence and/or absence of its specific target. The zsynthesized gold nanoparticles were able to differentiate between these two states by their characteristic surface plasmon resonance-based color change. Visual detection of METH in the low micromolar range is possible within minutes with the use of this method. This GNP-aptamer strategy helps in visual selective detection of micromolar quantities of METH within 30 min. The proposed method represents a generic GNP-based platform strategy [56].

S. Rouhani and S. Haghgoo synthesized a novel functionalized 1,8-naphthalimide derivative with a thiophene group sensitive to METH. Further, it was doped on the silica nanoparticle surface. A selective fluorescence “turn-on” nanosensor for methamphetamine based on a new 1,8-naphthalimide-thiophene derivative (NTS) and its immobilization on the surface of the nano-silica particles (NTS@SiO2).Fluorescence enhancement with high selectivity and sensitivity for METH was noticed in both solution and solid film. The NTS showed a significant fluorescence enhancement of (7.3-fold) in the presence of METH and was high after being attached to nanosilica particles (15.39-fold) due to the nanostructure effect.NTS@SiO2 showed high sensitivity to METH vapor in the solid state as well as in a solution. The emission of NTS@SiO2 in the film state exhibited a fast and reversible response to METH vapor with a linear correlation in the range of 6.7 μmol L^{-1} to 0.27 mmol L^{-1} [57].

3.5. Clonazepam Detection

Clonazepam is chemically 5-(2-chlorophenyl)-7-nitro-2,3-dihydro-1,4-benzodiazepine-2-one. It belongs to the benzodiazepine class of compounds having anxiolytic, anticonvulsant, muscle relaxant, amnestic, hypnotic and sedative properties. It is used for the treatment of epilepsy and panic disorders and also used for certain anxiety disorders and migraine. Buick Habibi et al. developed a simple and highly sensitive electrochemical method based on a silver nanoparticles/multi-walled carbon nanotubes nanocomposite modified glassy carbon electrode (AgNPs/MWCNTs/GCE) for the determination of clonazepam (CZP).They utilized AgNPs as it possesses high effective surface area and efficient electron transfer with inherent properties of MWCNTs. Also, AgNPs was synthesized on the functionalized MWCNTs without any additional reducing reagent. The cyclic voltammetry and differential pulse voltammetry (DPV) methods were used for the electrochemical studies and measurements. The results showed that the AgNPs/MWCNTs/GCE exhibited excellent electrocatalytic properties for clonazepam reduction reaction and the presence of the nanocomposite modifier on the GCE surface significantly affected the sensitivity of the electrochemical responses. The reduction peak currents showed a linear increase in the concentration range of 5.0×10^{-8} - 2.5×10^{-6} M with a detection limit of 6.0×10^{-9} M. The AgNPs/MWCNTs/GCE showed excellent stability, reproducibility and repeatability without any interference of other materials [58].

Rajeev Jain etal. also developed a voltammetric sensor based on a polyaniline/graphene oxide *modified* glassy carbon electrode (PANI/ GRO/GCE) for the ultrasensitive voltammetric detection of clonazepam. The fabricated PANI/GRO/ GCE sensor showed a high electrocatalytic response towards the reduction of CZPM in ethyl alcohol at pH 8.8 in BR buffer. The combined effect of PANI and GRO results in higher conductivity, surface area, and performance towards the reduction of clonazepam characterized by a square wave and cyclic voltammetry. The square wave cathodic current exhibited a linear response to clonazepam within the concentration range of 10 (ng mL^{-1}) to 550 (ng mL^{-1}) with a correlation coefficient (r2) of 0.9935, a detection limit (LOD) 4.63 ng mL^{-1} and a quantification limit (LOQ) 15.44 ng mL^{-1}. The reproducibility of 4.21% relative standard deviation (RSD) and a satisfactory recovery from 97.16% to 102.78% was obtained from the present method [59]. Similarly Mohamed Rizk et al. utilized carbon paste electrode modified with zinc oxide nanoparticles immobilized on multi-walled carbon nanotubes (MWCNTs/ZnONPs/CPE) for clonazepam detection in drug substance and combined drug products as well as in spiked human urine. The cyclic voltammograms showed well defined, irreversible cathodic peaks (reduction) at potential around −636 mV in Britton-Robinson (BR) buffer at pH 9.0. A good linear relationship was obtained between the peak current and the concentration within the range of 0.39–7.70 μg/mL.the LOD was found to be 0.173 μg/ml CLZ and LOQ was found to be 0.523 μg/ml [60].

3.6. Benzoylecgonine Detection

Tianxi Yang et al. used Au nanoparticles dotted magnetic nanocomposites (AMN) modified with inositol hexakisphosphate (IP6) as surface-enhanced Raman scattering (SERS) substrate to quickly monitor trace drug-related biomarkers in saliva and for on-site screening of trace drug biomarker in fingerprints. The presence of external magnet resulted in a huge SERS activity, which helped in the detection of the drug biomarkers in saliva from the U.S. Substance Abuse and Mental Health Services Administration. The limit of detection of benzoylecgonine in saliva was 29 ppb. The AMN-based SERS assay was demonstrated successfully using a portable Raman spectrometer, which could be useful for on-site detection and also helps in accurately differentiate between the smokers and drug addicts in near future [61].

3.7. ϒ-Hydroxybutyrate (GHB) Detection

ϒ-Hydroxybutyrate (GHB) is an ever more popular drug of abuse that acts as a central nervous depressant and causes stimulation, euphoria, anxiolysis or hypnosis, depending on the ad ministered dose. It has become one of the most dangerous illicit drugs. Low doses of the drug are used recreationally and also concerned in drug-facilitated sexual assaults. GHB is often substituted with its non-scheduled precursors g-butyrolactone(GBL) and 1,4-butanediol (BD), which are quickly metabolized into GHB in the body [62]. A chitosan/porous silicon biosensing platform for the detection of carboxylic acid-containing drugs in water was prepared and characterized. Optical interferometry measurements were performed on the chitosan/porous silicon substrate for the detection of GHB, a model molecule for illicit rape drug GHB, at concentrations compatible with forensic analysis[63]. A protein array chip for label-free optical detection of the individual components of a mixture of four common narcotics (cocaine, ecstasy, heroin, and amphetamine), using imaging surface plasmon resonance has been reported. The chip is produced by injecting a mixture of antibodies and letting them bind to narcotic analog coupled proteins already present in a predefined pattern on the supporting substrate. An indirect detection method, where antibodies are displaced from the surface upon recognition of their corresponding narcotics, was used to obtain the optical contrast and thus a detectable SPR and/or ellipsometric signal. Positive hits were routinely obtained for analyte concentrations of 50 pg μL^{-1} and the limit of detections was in the ranges of 0.5 pg μL^{-1} for heroin, 2.5 pg μL^{-1} for cocaine, and 5 pg μL^{-1} for the other two narcotics [64].

3.8. Multiplexed Drug Detection

M. W. J. Prins and their colleagues developed optomagnetic technology which allows high-sensitive one-step assays in blood serum/plasma and whole saliva. Drugs of abuse are detected at sub-nanogram per milliliter levels in a total assay time of 1 min. They developed a chip platform based on nanoparticles that are magnetically actuated and optically detected in a stationary sample fluid[65]. The drugs of abuse detected by developed nanochip platform. The first phase indicates the dried nanoparticles are redispersed and bind morphine molecules in the sample. In the second phase, nanoparticles are attracted towards the sensor surface with the lower magnet. In the third phase, weakly bound and unbound particles are removed with a magnetic force directed away from the surface using the upper magnet. The nanoparticles are coated with anti-morphine antibodies, BSA–morphine on the sensor surface. The data is obtained for detection of five drugs amphetamine (AMP), morphine (OPI), methamphetamine (MAMP), tetrahydrocannabinol (THC), and cocaine (COC) in 100% saliva. Results of a negative multiplexed assay (all five drugs at 0 ng mL^{-1}) and results of a positive multiplexed assay (AMP: 80 ng mL^{-1}, OPI: 48 ng mL^{-1}, MAMP: 80 ng mL^{-1}, THC: 800 ng mL^{-1} and COC: 48 ng mL^{-1}).

Recently, Chia-Wen Tsao and his colleagues reported nanostructured silicon (nSi) chip for the determination of abused drugs such as amphetamine, methamphetamine, codeine, morphine, and ketamine using nanostructured silicon surface-assisted laser desorption/ionization mass spectrometry (nSi-MS) [66]. Drug standard tests were applied to the chip platform to evaluate the nSi-MS performance, including detection sensitivity, the limit of detection, linearity, and repeatability. Real urine samples obtained from drug addict detainees were directly applied to the nSi chip for drug analysis and detected successfully from extremely small urine volumes (~10 μL), and a fast analysis procedure (<15 minutes).

CONCLUSION AND FUTURE PERSPECTIVES

Nanotechnologies have emerged as a significant role in the improvement of devices such as sensors, chips, and probes in different areas of forensic analysis. This chapter shows clearly the potentiality that nano-enabled sensors can be used as rapid, reliable and affordable analytical tools in this branch of modern analytical chemistry that is characterized by the legal context affecting both the work implementation and the results, and the variety of analytes able to be investigated. As it has been intuiting through some selected examples, a great diversity of interesting nanosensors designs has been developed. Nevertheless, the routine use nanodevices in the forensic practice are still very limited, probably as a consequence of that many of the reported developments were not applied to samples with the grade of complexity that is very essential. Therefore, major research efforts must be made in the near future to ensure the preparation of nanodevice better equipped for these samples, in order to enhance the robustness and reliability necessary for real applicability to this field. The sensitivity and performance of the device are being improved by using nanomaterials for their construction. The use of nanomaterials has allowed the introduction of many new signal transduction technologies in devices. Nanodevices offer new potential capabilities such as miniaturization to nanoscale size, higher sensitivities, and reduced cost of production. This will greatly improve functionality, by providing devices that are small, portable, and easy to use in day to day activities. This technology is going to revolutionize as future armor in the fields of virtopsy, crime scene investigation, identification, forensic ballistics, and toxicology.

While writing this chapter, we have noticed that only a few drugs have been preferentially studied by researchers since not much work has been carried out on these drugs. Literature reveals that codeine, morphine, cocaine, etc. have been the most abused drugs around the world. We hope that after reading this critical review, researchers around the world will be motivated to enter into unknown territories of 'not so common' abusing drugs. There are a plethora of drugs whose nanotechnology is yet to be studied. We also noted that the preferred use of certain nanomaterials, mainly carbon Nanomaterial-based electrochemical sensing of drugs nanotubes and other carbon species as well as gold nanoparticles (NPs) to prepare chemically modified electrodes, whilst nanoparticles made from silver, platinum, palladium, or metal oxides have not been employed vastly. Employing such nanoparticles as modifiers will predictably form a highly interesting topic for future research in the field of the nanomaterial-based abusing drug detection device.

Acknowledgments

We gratefully acknowledge the funding from SERB, India under Young Scientist Start-up Grant (YSS/2015/000258).

References

[1] Nasir Uddin, M. (2014). An Overview on Total Analytical Methods for the Detection of 1,4-Benzodiazepines. *Pharmaceutica Analytica Acta*, 05(06).

[2] Zhai, D., Tan, Y., Xu, W. and Chang, Y. (2014). Development of a fluorescent sensor for illicit date rape drug GHB. *Chemical Communications*, 50(22), p.2904.

[3] Sanghavi, B., Wolfbeis, O., Hirsch, T. and Swami, N. (2014). Nanomaterial-based electrochemical sensing of neurological drugs and neurotransmitters. *Microchimica Acta*, 182(1-2), pp.1-41.

[4] Zhai, D., Tan, Y., Xu, W. and Chang, Y. (2014). Development of a fluorescent sensor for illicit date rape drug GHB. *Chemical Communications,* 50(22), p.2904.

[5] Lendoiro, E., Quintela, Ó., de Castro, A., Cruz, A., López-Rivadulla, M. and Concheiro, M., 2012. Target screening and confirmation of 35 licit and illicit drugs and metabolites in hair by LC–MSMS. *Forensic Science International*, 217(1-3), pp.207-215.

[6] Pedersen, A., Dalsgaard, P., Rode, A., Rasmussen, B., Müller, I., Johansen, S. and Linnet, K., 2013. Screening for illicit and medicinal drugs in whole blood using fully automated SPE and ultra-high-performance liquid chromatography with TOF-MS with data-independent acquisition. *Journal of Separation Science*, 36(13), pp.2081-2089.

[7] Burns, L. (2014). World Drug Report 2013 By United Nations Office on Drugs and Crime New York: United Nations, 2013ISBN: 978-92-1-056168-6, 151 pp. Grey literature. *Drug and Alcohol Review*, 33(2), pp.216-216.

[8] Klein, R. and Hays, P. (2002). Detection and analysis of drugs of forensic interest, 1992-2001. Alexandria, VA: U.S. *Drug Enforcement Administration, Office of Forensic Sciences,* Laboratory Support Section.

[9] Yáñez-Sedeño, P., Agüí, L., Villalonga, R. and Pingarrón, J. (2014). Biosensors in forensic analysis. A review. *Analytica Chimica Acta,* 823, pp.1-19.

[10] Leon, M. B., W. Diana, S. Robert, Current trends in nanobiosensor technology. *Nanomed Nanobiotechnol*. 3 (2011) 229–246.

[11] Lad, A., Pandya, A. and Agrawal, Y. (2016). Overview of nano-enabled screening of drug-facilitated crime: A promising tool in forensic investigation. *TrAC Trends in Analytical Chemistry*, 80, pp.458-470.

[12] Pandya, A. and Shanker, R. (2017). Smartphone Based Test Strip Platform for Monitoring of Fluoride Ions in Water Samples. *Advanced Science, Engineering and Medicine*, 9(8), pp.619-627.

[13] Pandya, A., Sutariya, P., Lodha, A. and Menon, S. (2013). A novel calix[4]arenethiol functionalized silver nanoprobe for selective recognition of ferric ion with nanomolar sensitivity via DLS selectivity in human biological fluid. *Nanoscale*, 5(6), p.2364.

[14] Pandya, A., Sutariya, P. and Menon, S. (2013). A non enzymatic glucose biosensor based on an ultrasensitive calix[4]arene functionalized boronic acid gold nanoprobe for sensing in human blood serum. *The Analyst*, 138(8), p.2483.

[15] Chen, X., Cui, D., Liu, C. and Cai, H. (2006). Microfluidic Biochip for Blood Cell Lysis. *Chinese Journal of Analytical Chemistry,* 34(11), pp.1656-1660.

[16] Du, W., Xu, Z., Ma, X., Song, L. and Schneider, E. (2003). Biochip as a potential platform of serological interferon α2b antibody assay. *Journal of Biotechnology*, 106(1), pp.87-100.

[17] Dave, H., Bhandari, N., Mod, B., Shukla, R. and Pandya, A. (2018). Multifunctional Silver-Cellulose Nanocomposite as a Promising Plasmonic Sensing Platform. *Journal of Nanoscience and Nanotechnology*, 18(8), pp.5461-5469.

[18] Campàs, M. and Katakis, I. (2004). DNA biochip arraying, detection and amplification strategies. *TrAC Trends in Analytical Chemistry,* 23(1), pp.49-62.

[19] Xu, K., Huang, J., Ye, Z., Ying, Y. and Li, Y. (2009). Recent Development of Nano-Materials Used in DNA Biosensors. *Sensors,* 9(7), pp.5534-5557.

[20] Jain, K. (2005). Nanotechnology in clinical laboratory diagnostics. *Clinica Chimica Acta*, 358(1-2), pp.37-54.

[21] L. Huang, Z. Peng, Y. Guo, A. L. Porter, Identifying the emerging roles of nanoparticles in biosensors. *J of Buss. Chem.* 7 (2010) 15-30.

[22] Vikesland, P. and Wigginton, K. (2010). Nanomaterial Enabled Biosensors for Pathogen Monitoring - A Review. *Environmental Science& Technology*, 44(10), pp.3656-3669.

[23] El-Ansary, A. (2010). Nanoparticles as biochemical sensors. *Nanotechnology, Science and Applications*, p.65.

[24] Khanna, V. (2008). New-generation nano-engineered biosensors, enabling nanotechnologies and nanomaterials. *Sensor Review*, 28(1), pp.39-45.

[25] Kim, T., Park, I., Nah, J., Choi, Y. and Cho, C. (2004). Galactosylated chitosan/DNA nanoparticles prepared using water-soluble chitosan as a gene carrier. *Biomaterials,* 25(17), pp.3783-3792.

[26] Kumar, V., Kumar, P., Pournara, A., Vellingiri, K. and Kim, K. (2018). Nanomaterials for the sensing of narcotics: Challenges and opportunities. *TrAC Trends in Analytical Chemistry*, 106, pp.84-115.

[27] L. L. Brunton, J. S. Lazo, K. L. Parker, *Goodman & Gilman's The Pharmacological Basis Of Therapeutics,* Eleventh Ed., McGraw-Hill 2006.

[28] H. P. Rang, M. M. Dale, J. M. Ritter, R. J. Flower, G. Henderson, *Rang and Dale's Pharmacology*, seventh Ed., Elsevier, Churchill Livingstone 2012.

[29] Farrell, S. and Golan, D. (2008). *Principles of pharmacology.* Philadelphia: Wolters Kluwer Health/ Lippincott Williams & Wilkins.

[30] Niu, X., Huang, L., Zhao, J., Yin, M., Luo, D. and Yang, Y. (2016). An ultrasensitive aptamer biosensor for the detection of codeine based on a Au nanoparticle/polyamidoaminedendrimer-modified screen-printed carbon electrode. *Analytical Methods*, 8(5), pp.1091-1095.

[31] Babaei, A., Afrasiabi, M. and Shabanian, M. (2017). Application of multivariate optimization method in nanomolar simultaneous determination of morphine and codeine in the presence of uric acid using a glassy carbon electrode modified with a hydroxyapatite-Fe3O4 nanoparticle/multiwalled carbon nanotubes composite. *Journal of the Iranian Chemical Society*, 14(11), pp.2305-2317.

[32] Afkhami, A., Khoshsafar, H., Bagheri, H. and Madrakian, T. (2014). Facile simultaneous electrochemical determination of codeine and acetaminophen in pharmaceutical samples and biological fluids by graphene–CoFe2O4 nancomposite modified carbon paste electrode. *Sensors and Actuators B: Chemical*, 203, pp.909-918.

[33] Taei, M., Salavati, H., Hasanpour, F., Habibollahi, S. and Baghlani, H. (2016). Simultaneous determination of ascorbic acid, acetaminophen and codeine based on multi-walled carbon nanotubes modified with magnetic nanoparticles paste electrode. *Materials Science and Engineering*: C, 69, pp.1-11.

[34] Mashhadizadeh, M. and Rasouli, F. (2014). Design of a New Carbon Paste Electrode Modified with TiO_2Nanoparticles to Use in an Electrochemical Study of Codeine and Simultaneous Determination of Codeine and Acetaminophen in Human Plasma Serum Samples. *Electroanalysis*, 26(9), pp.2033-2042.

[35] Ensafi, A., Ahmadi, N., Rezaei, B. and Abarghoui, M. (2015). A new electrochemical sensor for the simultaneous determination of acetaminophen and codeine based on porous silicon/palladium nanostructure. *Talanta,* 134, pp.745-753.

[36] Ali Babaei, M. Babazadeh and H. R. Momeni (2011) "A Sensor for Simultaneous Determination of Dopamine and Morphine in Biological Samples Using a Multi-Walled Carbon Nanotube/ Chitosan Composite Modified Glassy Carbon Electrode" *Int. J. Electrochem. Sci.* 6:1382 – 1395.

[37] Ya, Y., Xiaoshu, W., Qing, D., Lin, J. and Yifeng, T. (2015). Label-free immunosensor for morphine based on the electrochemiluminescence of luminol on indium–tin oxide coated glass functionalized with gold nanoparticles. *Analytical Methods*, 7(11), pp.4502-4507.

[38] Atta, N., Galal, A. and Azab, S. (2011). Determination of morphine at gold nanoparticles/Nafion® carbon paste modified sensor electrode. *The Analyst*, 136(22), p.4682.

[39] Nada F. Atta, Ahmed Galal, Anwar A. Wassel and Asmaa H. Ibrahim (2012) Sensitive Electrochemical Determination of Morphine Using Gold Nanoparticles–Ferrocene Modified Carbon Paste Electrode. *Int. J. Electrochem. Sci.* 7:10501 – 10518.

[40] Nada F. Atta, Ahmed Galal and Ekram H. El-Ads (2014) Smart Electrochemical Morphine Sensor Using poly(3,4-ethylene-dioxythiophene)/Gold-nanoparticles Composite in Presence of Surfactant. *Int. J. Electrochem. Sci.* 9: 2113 – 2131.

[41] Gandhi, S., Banga, I., Maurya, P. and Eremin, S. (2018). A gold nanoparticle-single-chain fragment variable antibody as an immunoprobe for rapid detection of morphine by dipstick. *RSC Advances*, 8(3), pp.1511-1518.

[42] Eissa, S. and Zourob, M. (2017). Competitive voltammetric morphine immunosensor using a gold nanoparticle decorated graphene electrode. *MicrochimicaActa*, 184(7), pp.2281-2289.

[43] Dehdashtian, S., Gholivand, M., Shamsipur, M. and kariminia, S. (2016). Construction of a sensitive and selective sensor for morphine using chitosan coated Fe 3 O 4 magnetic nanoparticle as a modifier. *Materials Science and Engineering*: C, 58, pp.53-59.

[44] Bagheri, H., Khoshsafar, H., Afkhami, A. and Amidi, S. (2016). Sensitive and simple simultaneous determination of morphine and codeine using a Zn2SnO4 nanoparticle/graphene composite modified electrochemical sensor. *New Journal of Chemistry*, 40(8), pp.7102-7112.

[45] Fei, W., Chen, F., Sun, L., Li, Q., Yang, J. and Wu, Y. (2013). Ultrasensitive electrochemiluminescent immunoassay for morphine using a gold electrode modified with CdS quantum dots, polyamidoamine, and gold nanoparticles. *MicrochimicaActa*, 181(3-4), pp.419-425.

[46] Ensafi, A., Izadi, M., Rezaei, B. and Karimi-Maleh, H. (2012). N-hexyl-3-methylimidazolium hexafluoro phosphate/multiwall carbon nanotubes paste electrode as a biosensor for voltammetric detection of morphine. *Journal of Molecular Liquids*, 174, pp.42-47.

[47] Masteri-Farahani, M., Khademabbasi, K. and Mollatayefeh, N. (2018). A selective morphine nanosensor derived from functionalized CdS quantum dots. *Materials Letters*, 228, pp.68-71.

[48] Roushani, M. and Shahdost-fard, F. (2015). A novel ultrasensitive aptasensor based on silver nanoparticles measured via enhanced voltammetric response of electrochemical reduction of riboflavin as redox probe for cocaine detection. *Sensors and Actuators B: Chemical*, 207, pp.764-771.

[49] Taghdisi, S., Danesh, N., Emrani, A., Ramezani, M. and Abnous, K. (2015). A novel electrochemical aptasensor based on single-walled carbon nanotubes, gold electrode and complimentary strand of aptamer for ultrasensitive detection of cocaine. *Biosensors and Bioelectronics*, 73, pp.245-250.

[50] Emrani, A., Danesh, N., Ramezani, M., Taghdisi, S. and Abnous, K. (2016). A novel fluorescent aptasensor based on hairpin structure of complementary strand of aptamer and nanoparticles as a signal amplification approach for ultrasensitive detection of cocaine. *Biosensors and Bioelectronics*, 79, pp.288-293.

[51] Li, K., Qin, W., Li, F., Zhao, X., Jiang, B., Wang, K., Deng, S., Fan, C. and Li, D. (2013). Nanoplasmonic Imaging of Latent Fingerprints and Identification of Cocaine. *Angewandte Chemie International Edition*, 52(44), pp.11542-11545.

[52] Hashemi, P., Bagheri, H., Afkhami, A., Ardakani, Y. and Madrakian, T. (2017). Fabrication of a novel aptasensor based on three-dimensional reduced graphene oxide/polyaniline/gold nanoparticle composite as a novel platform for high sensitive and specific cocaine detection. *Analytica Chimica Acta*, 996, pp.10-19.

[53] Chen, Z., Tan, Y., Xu, K., Zhang, L., Qiu, B., Guo, L., Lin, Z. and Chen, G. (2016). Stimulus-response mesoporous silica nanoparticle-based chemiluminescence biosensor for cocaine determination. *Biosensors and Bioelectronics*, 75, pp.8-14.

[54] Zhimin Cai, Zhijie Lin, Xiaomei Chen, Tiantian Jia, Ping Yu, and Xi Chen (2010) "Electrochemiluminescence detection of methamphetamine based on a $Ru(bpy)_3^{2+}$- doped silica nanoparticles/Nafi on composite film modified electrode" *Luminescence* 24: 367–372.

[55] Mao, K., Yang, Z., Li, J., Zhou, X., Li, X. and Hu, J. (2017). A novel colorimetric biosensor based on non-aggregated Au@Ag core–shell nanoparticles for methamphetamine and cocaine detection. *Talanta,* 175, pp.338-346.

[56] Melina Yarbakht & Maryam Nikkhah (2016) Unmodified gold nanoparticles as a colorimetric probe for visual methamphetamine detection, *Journal of Experimental Nanoscience* 11: 593-601.

[57] Rouhani, S. and Haghgoo, S. (2015). A novel fluorescence nanosensor based on 1,8-naphthalimide-thiophene doped silica nanoparticles, and its application to the determination of

methamphetamine. *Sensors and Actuators B: Chemical*, 209, pp.957-965.

[58] Habibi, B. and Jahanbakhshi, M. (2014). Silver nanoparticles/multi walled carbon nanotubes nanocomposite modified electrode: Voltammetric determination of clonazepam. *ElectrochimicaActa*, 118, pp.10-17.

[59] Jain, R., Sinha, A., Kumari, N. and Khan, A. (2016). A polyaniline/graphene oxide nanocomposite as a voltammetric sensor for electroanalytical detection of clonazepam. *Analytical Methods*, 8(15), pp.3034-3045.

[60] Rizk, M., Taha, E., El-Alamin, M., Hendawy, H. and Sayed, Y. (2018). Highly Sensitive Carbon Based Sensors Using Zinc Oxide Nanoparticles Immobilized Multiwalled Carbon Nanotubes for Simultaneous Determination of Desvenlafaxine Succinate and Clonazepam. *Journal of The Electrochemical Society*, 165(7), pp.H333-H341.

[61] Yang, T., Guo, X., Wang, H., Fu, S., wen, Y. and Yang, H. (2015). Magnetically optimized SERS assay for rapid detection of trace drug-related biomarkers in saliva and fingerprints. *Biosensors and Bioelectronics,* 68, pp.350-357.

[62] A. Kankaanpaa, R. Liukkonen, K. Ariniemi, . Determination of γ-hydroxybutyrate (GHB) and its precursors in blood and urine samples: A salting-out approach. *Foren. Sci. Intern.* 170 (2007) 133-138.

[63] B. Sciacca, E. Secret, S. Pace, P. González, F. Geobaldo, F. Quignard, F. Cunin. Chitosan-functionalized porous silicon optical transducer for the detection of carboxylic acid-containing drugs in water. *Journal of Materials Science* 21 (2011) 2294-2302.

[64] G. Klenkar, B. Liedberg, A microarray chip for label-free detection of narcotics. *Analy. And Bioana. Chem.* 391 (2008) 1679-1688.

[65] D. M. Bruls, T. H. Evers, J. A. H. Kahlman, P. J. W. van Lankvelt, M. Ovsyanko, E. G. M. Pelssers, J. J. H. B. Schleipen, F. K. de Theije, C. A. Verschuren, T. van der Wijk, J. B. A. van Zon, W. U. Dittmer, A. H. J. Immink, J. H. Nieuwenhuis and M. W. J. Prins.

Rapid integrated biosensor for multiplexed immunoassays based on actuated magnetic nanoparticles. *Lab Chip* 9 (2009) 3504–3510.

[66] C. Yu-Che, C. Kun-Han, W. Juo-Shin, H. Wen-Liu, C. Chih-Cheng, C. Wen-Yih, and T. Chia-Wen, Rapid analysis of abused drugs using nanostructured silicon surface assisted laser desorption/ionization mass spectrometry. *Analyst* 137 (2012) 654-661.

ABOUT THE EDITORS

Ritesh Kumar Shukla, PhD

Dr. Ritesh Kumar Shukla is working as Assistant Professor at Biological and Life Sciences Division, School of Arts and Sciences, Ahmedabad University, Ahmedabad, India. He did his master's in Forensic Sciences from Dr. BR Ambedkar University, Agra (2005) and PhD in Toxicology from CSIR-Indian Institute of Toxicology Research, Lucknow/Jamia Hamdard, New Delhi (2013). His PhD work was focused on understanding the adverse effect of engineered nanoparticles through in vitro and in vivo approaches. He has been involved in developing various methods pertaining to techniques in the field of nanotoxicology. He has co-authored more than 30 research articles in international peer reviewed journals and 3 book chapters. He has received the "Early Career Research Award" from the Science and Engineering Research Board, Department of Science and Technology, Government of India in 2016 to develop R&D in the field of Forensic Biology/Nano-Biotechnology in India. He has research expertise in diversified field of Toxicology (Genetic toxicology, Nanotoxicology and radiation toxicology) and Forensic Science (DNA Forensics). Currently his research focused on to develop paper based affordable device for forensic applications. He is the member of prestigious scientific society in the field of forensic like Silk Road Forensic Consortium, China (Founded by Professor Henry Lee) and International society of Forensic Genetics (ISFG).

Before joining Ahmedabad University, he also served as Assistant Professor at the Institute of Forensic Science, Nagpur (2011-12) and Guru Ghasidas Vishwavidyalaya, a Central University, Bilaspur (2012-13).

Alok Pandya, PhD

Dr. Alok Pandya is an Assistant Professor in Department of Physical Sciences, Institute of Advanced Research, Gujarat, India. He has worked as Young Scientist (2016-17) and Post-Doctoral Research Associate (2013-16) at Division of biological Sciences, Ahmedabad University, Gujarat, India. He completed his PhD from Gujarat University in 2013 and awarded with "GSA-CHARUSAT Best Thesis Award". He has worked on development of novel synthetic methodologies of supra-molecular nanoassembly for applications in sensing and catalysis during PhD program. He has also developed various nanoassemblies for the detection of illicit drugs in forensic context. In his postdoctoral research, he was involved in the project on development of point of care diagnostic for cardiac disease using paper microfluidic. He has worked on synthesis and development of nanobiosensors targeting to different cardiac specific protein biomarker which could translate into paper based LFIA based nanodevice, synthesis of new class of nanomaterials such as magnesium nanocomplex (MgNCs) with biological support for imaging applications. He has published more than 40 articles in high repute international journals and published 5 books/book chapters. His research interests span an interdisciplinary research combines nano-chemistry, nano-biology and nano-fabrication disciplines enabling the development of multifunctional nanoplatform focusing.

INDEX

μ

μ receptor, 261
μ1 and μ2 receptor, 261

#

2,4,6-trinitrotoluene (TNT), 116, 120, 121, 122, 123, 124, 125, 126, 127, 128, 130, 137, 138, 139, 140, 141
5-HT receptor activation, 260
5-HT2, 260

A

A document, 236
ABAcard, 105, 106, 113
abuse, 108, 228, 242, 256, 266, 275
abused drugs, 256, 260, 280, 281, 290
acetaminophen, 262, 271
additives, 190, 192
admissible, 240, 241
adsorption, 73, 124, 133, 185, 266
AFLP, 146, 150, 159, 160, 185
aggregation, 36, 93, 94, 122, 202, 203, 274, 275
agonist, 260, 261
alfentanyl, 261
aliphatic nitro, 116
alkaloids, 229, 231, 265
aluminium, 39, 71
amendments, 246
amine, 59, 62, 64, 67, 122, 127, 133, 139
amino, 27, 52, 64, 65, 67, 77, 78, 81, 127, 146, 196, 260, 270
amino acid, 27, 52, 64, 65, 67, 77, 78, 81, 146, 196
ammonium, 116, 133, 167, 245
amphetamines, 261
aniline, 123, 137, 138
annealing, 153, 154, 157
antibody, 88, 90, 91, 106, 123, 125, 213, 266, 269, 271, 272
anti-conterfeiting, 44
antigen, 90, 91, 97, 106
applications and forensic science, 1
aptamer, 31, 66, 213, 257, 262, 273, 274, 275, 276, 285, 287, 288
aqueous solutions, 128, 129, 262
aromatic compounds, 122, 126
arsenic, 111, 192, 210, 257
ascorbic acid, 188, 263, 264, 268, 271
assault, 107, 108, 161, 256

assay, 93, 102, 104, 106, 107, 113, 123, 137, 198, 212, 213, 270, 273, 278, 280, 284, 289
atoms, 2, 4, 6, 11, 14, 16, 19, 34, 36, 40
atropine, 261, 268
Au nanoparticles, 66, 123, 185

B

bacteria, 43, 158, 187, 191, 201, 203, 210, 213
barbiturates, 260
benzodiazapines, 260
bind with serotonin in nerve ending, 260
bioassay, 128, 276
biocompatibility, 258, 273
biological, vii, viii, 6, 19, 27, 42, 43, 82, 85, 86, 87, 88, 90, 91, 92, 94, 96, 99, 100, 101, 102, 103, 110, 111, 126, 128, 140, 145, 146, 147, 149, 150, 158, 161, 165, 167, 174, 183, 184, 187, 190, 198, 199, 200, 201, 213, 227, 256, 258, 260, 262, 264, 266, 272, 283, 285, 286, 291, 292
biological fluids, 85, 99, 103, 260, 262, 266
biological processes, 86, 190
biological samples, 102, 103, 146, 147, 167, 257, 264, 266
biomarkers, 103, 182
biomolecular, 90, 95, 126, 140
biomolecules, 82, 259
biosensors, xi, 42, 48, 86, 87, 88, 105, 111, 146, 167, 177, 182, 187, 211, 212, 257, 258
bisphenol, 192, 206
black tea, 196, 207
blood, 26, 31, 53, 66, 67, 71, 73, 77, 82, 83, 85, 96, 97, 98, 99, 100, 101, 102, 103, 104, 105, 107, 108, 109, 112, 113, 145, 146, 147, 165, 169, 179, 180, 182, 185, 257, 259, 262, 265, 269, 271, 280, 283, 284, 289
blood group, 182, 185
body fluid, vii, 27, 85, 86, 96, 97, 98, 100, 101, 102, 103, 104, 108, 109, 110, 112, 113, 162, 163
bonding, 22, 94
bone, 26, 45, 146, 163
bottom-up, 8, 9, 11
bulk materials, 6, 8, 23, 36, 120

C

cadmium, 25, 64, 192
caffeine, 262, 264
calcium, 188, 193, 219
cancer, 19, 140, 190, 193, 212
cannabis, 228, 229, 230, 256, 260, 261
capillary, 22, 103, 129, 164, 198, 209, 266
carbon, 4, 29, 39, 40, 41, 44, 48, 90, 108, 119, 120, 124, 126, 127, 131, 134, 136, 139, 167, 178, 185, 246, 262, 264, 267, 270, 272, 273, 274, 285
carbon nanotube, 4, 44, 120, 124, 126, 127, 131, 134, 139, 141, 142, 143, 167, 185, 246, 262, 264, 267, 272, 273, 277, 278, 285, 286, 287, 289
carboxylic acid, 62, 64, 65, 67
carotid chemoreceptor inhibitor, 261
catalysis, 95, 121
catalytic properties, 166, 258
CB1, 261
CB1-receptors, 261
CB2 agonist, 261
central nervous system, 86, 265, 272, 275
ceramic, 69, 100
characterization, viii, 1, 2, 5, 14, 30, 37, 86, 180, 207, 208, 211
charas, 229, 234
chemical, 3, 6, 7, 9, 10, 11, 12, 13, 19, 22, 24, 29, 30, 31, 36, 38, 42, 43, 48, 50, 52, 54, 55, 56, 57, 67, 68, 72, 77, 81, 83, 86, 87, 89, 90, 91, 92, 93, 95, 96, 99, 100,

105, 110, 111, 115, 116, 119, 120, 125, 126, 128, 139, 140, 185, 187, 188, 189, 190, 191, 193, 194, 195, 196, 198, 199, 200, 201, 203, 204, 205, 206, 212, 217, 219, 220, 223, 225, 227, 234, 244, 247, 257, 258, 259, 282, 285, 287, 289
chemical etching, 30, 95
chemical properties, 3, 50, 72, 217, 219, 223, 259
chemical vapor deposition, 13, 125
chemiresistive, 125, 139
children, 171, 193, 206, 256
chitosan, 39, 47, 69, 262, 267, 270, 273
chromatographic, 221, 257
chromatography, 120, 136, 196, 207, 210, 221, 266, 273, 275
chromosome, 157, 161, 170, 179
citizens, 135, 171, 174, 248
clarity, 61, 66
classes, 42, 116, 189
classification, 101, 197
clustering, 101, 212
clusters, 2, 30, 34, 35, 36, 201, 203
CNT, 90, 92, 140
Coca derivative, 229
Coca leaf, 229
cocaine, 229, 256, 260, 261, 272, 273, 274, 275, 279, 280, 281, 287, 288
codeine, 231, 260, 261, 262, 263, 264, 265, 280, 281, 285, 286, 287
coding, 82, 157, 158, 166, 179, 225
coffee, 193, 197, 208
color, 26, 42, 56, 60, 64, 93, 94, 107, 108, 122, 136, 194, 201, 202, 269
colorimetric, 43, 65, 77, 81, 92, 93, 94, 111, 112, 122, 137, 201, 202, 203, 212, 275, 288
commercial, vii, 17, 19, 40, 41, 181, 207, 208
complexity, 4, 52, 73, 204
composites, 127, 210
composition, 11, 16, 17, 19, 24, 54, 72, 77, 90, 95, 96, 107, 118, 158, 207, 219, 220, 224
compounds, 38, 41, 52, 53, 56, 61, 78, 116, 117, 122, 126, 132, 134, 136, 190, 191, 192, 197, 247, 265
conductivity, 5, 120, 124, 132, 273
conference, 175, 185
confidentiality, 172, 173, 185
configuration, 22, 91, 123
confinement, 201, 236
consciousness, 228
consent, 173, 256
conservation, 73, 175, 218, 219, 224, 225
constituents, 13, 23, 52, 53
constitution bench, 241
construction, 116, 117, 158, 242, 259
consumption, 193, 194, 206, 228
contaminant, 25, 191, 197, 200, 202, 203
contaminated soil, 127, 134
contamination, 10, 111, 120, 134, 190, 191, 192, 194, 195, 204, 205, 206, 209, 211
contract, 237
conviction, 233, 234, 235, 238, 239
copper, 36, 40, 71, 134, 167, 180, 201, 212, 262
correlation, 27, 166, 268, 276
cost, viii, 27, 42, 50, 56, 74, 103, 118, 123, 127, 167, 196, 198, 199, 204, 255, 259, 264, 267
cotinine, 41, 47
counterfeiting, 38, 44, 221, 222, 225
creatinine, 108, 111
crimes, 42, 110, 135, 174, 235, 247, 275
criminal activity, 225, 246
criminals, 39, 50
crystalline, 15, 16, 203
crystals, 14, 22, 35, 39
cultural heritage, 216, 219, 224, 225
currency, 44, 55, 188, 221
cycles, 154, 157, 164, 171
cyclic voltammetry, 265, 272

cyclodextrins, 59, 65
cystine, 27, 32
cytosine, 146, 163

D

deaths, 43, 190, 193, 248, 256
decay, 83, 218
defects, 16, 264
deficiency, 188, 272
deformation, 90, 121
degradation, 162, 163, 216, 219
degradation process, 216, 219
denaturation, 153, 154
deposition, 12, 13, 30, 39, 53, 54, 57, 58, 59, 60, 65, 66, 75, 76, 77, 78, 79, 81, 125
derivatives, 229, 230, 260, 265
desorption, 38, 40, 41, 133
destruction, 99, 118
detectable, 86, 100, 106, 119, 124, 154
detection system, 34, 43, 44, 187
detection techniques, 78, 201
detonator, 116, 117
diagnostics, 35, 105, 112, 113, 158, 212, 284
dielectric, 93, 121, 212
diet, 51, 53
digestion, 149, 151, 152, 153, 154, 155, 161
direct evidence, 239, 246
disaster, 162, 175
disclosure, 173, 234
discrimination, 98, 101, 108, 112, 240
diseases, 146, 147, 158, 171, 182, 188, 189, 193
displacement, 16, 89, 124
diversity, 42, 158, 161, 257
DNA, vi, 24, 26, 31, 32, 42, 43, 48, 64, 90, 91, 96, 97, 99, 100, 101, 104, 107, 109, 110, 126, 134, 135, 141, 145, 146, 147, 148, 149, 150, 152, 153, 154, 155, 156, 157, 158, 159, 160, 161, 162, 163, 164, 165, 166, 167, 168, 169, 170, 171, 172, 173, 174,175, 176, 177, 178, 179, 180, 181, 182, 183, 184, 185, 195, 198, 199, 201, 202, 206, 207, 210, 211, 212, 246, 257, 273, 276, 284, 292
DNA fingerprinting, 145, 146, 147, 148, 149, 150, 159, 162, 164, 167, 168, 169, 170, 171, 174, 175, 178, 181, 182, 183, 184, 185
DNA polymerase, 153, 154
DNA testing, 173, 177
document examination, 215, 216, 217, 218, 224
document preservation, vii, viii, 216
dogs, 106, 118
donors, 99, 161, 164, 165
dopamine receptor antagonist, 261
dopamine transporter, 261
doxapram, 261
drug abuse, 228, 266
drug delivery, 5, 82
drug-facilitated crime, 256
drugs, vii, 33, 40, 44, 136, 188, 228, 229, 230, 232, 235, 246, 251, 252, 256, 257, 258, 259, 260, 262, 266, 276, 279, 280, 281, 282, 283, 289
dyes, 38, 39, 42, 203, 220, 221

E

effect on α and β receptor, 261
electrochemical, 11, 30, 42, 48, 88, 90, 91, 119, 120, 134, 136, 137, 138, 139, 141, 143, 183, 199, 203, 258, 262, 264, 265, 267, 268, 270, 272, 273, 274, 277, 282, 285, 286, 287, 289
electrochemical impedance, 268, 272, 275
electrochemical sensors, 42
electrochemistry, 268, 272
electrode surface, 42, 124, 264, 267, 270

electrodes, 123, 124, 136, 211, 262, 264, 270
electromagnetic, 14, 20, 23, 93
electromagnetic waves, 14, 23
electron, 12, 16, 17, 18, 19, 20, 24, 30, 36, 57, 90, 91, 92, 131, 132, 136, 196, 201, 207, 221, 262, 264, 267, 268, 270, 272, 273
electron microscopy, 17, 18, 30, 196
electronic structure, 19, 29
electrophoresis, 103, 149, 152, 153, 154, 157, 159, 160, 161, 164, 167, 182, 196, 198, 209, 266
electrospun, 132, 142
emission, 17, 20, 29, 61, 63, 67, 70, 71, 72, 73, 83, 203, 274
endonuclease, 150, 151, 153, 154, 155, 161, 182
energy, 5, 8, 10, 11, 17, 24, 30, 35, 40, 41, 44, 91, 116, 121, 123, 188, 203, 221, 272
energy transfer, 121, 123, 272
enforcement, 135, 247, 249, 250
environment, vii, 25, 74, 87, 93, 99, 123, 190, 259
environmental conditions, 53, 99, 189
environmental health, 250
environmental impact, 217, 219
enzymatic assay, 120
enzyme, 88, 89, 90, 91, 105, 107, 108, 150, 151, 153, 154, 155, 159, 160, 161, 188, 189, 198, 209, 213, 257, 260, 273
epidemic, 189, 206
epithelial cells, 107, 163
ester, 64, 69, 70, 105, 136, 229
etching, 10, 94
ethanol, 260, 271
ethical issues, 181, 227
ethylene, 116, 268
europium, 62, 80
evaporation, 12, 123
evidence, 24, 25, 37, 39, 49, 50, 53, 96, 99, 100, 101, 103, 108, 117, 135, 146, 149, 150, 174, 176, 181, 216, 227, 233, 234, 235, 237, 238, 239, 240, 241, 242, 245, 246, 247, 250
evidence act, 228, 234, 239, 240, 241
evolution, 94, 158, 179, 183
examinations, 106, 217, 220, 221
excitation, 69, 71, 91, 99, 201, 203
execution, 238, 243
ex-fentanyl, 261
expert evidence, 239
explosive, vii, viii, 33, 35, 43, 44, 55, 115, 116, 117, 118, 119, 120, 122, 125, 126, 127, 130, 131, 132, 133, 134, 135, 136, 137, 138, 139, 140, 141, 142, 143, 190, 242, 243, 244, 245, 246, 247, 253
explosive substances act, 242, 243, 244, 245, 253
exposure, 123, 134, 190, 206, 224, 250
extraction, 32, 45, 146, 150, 161, 164, 167, 175, 176, 181, 185, 196, 197, 199, 210

F

fabrication, 9, 10, 13, 168, 263, 267
false positive, 101, 102, 123
families, 117, 248
fibers, 26, 215, 216, 219, 221
fingerprint, vii, 24, 25, 31, 38, 39, 47, 49, 50, 51, 52, 53, 54, 55, 56, 57, 58, 59, 60, 61, 62, 63, 64, 65, 66, 69, 70, 71, 72, 73, 74, 75, 76, 77, 78, 79, 80, 81, 82, 83, 100, 112, 117, 146, 147, 150, 172, 178, 247, 278, 289
fluid, 85, 86, 96, 97, 98, 99, 100, 101, 102, 103, 104, 107, 108, 109, 110, 112, 113
fluorescence, 41, 43, 61, 62, 66, 69, 70, 71, 122, 123, 127, 130, 132, 136, 203, 213, 220, 266, 272, 274
food, vii, viii, 111, 175, 187, 188, 189, 190, 191, 192, 193, 194, 195, 196, 197, 198,

199, 203, 205, 206, 207, 208, 209, 210, 213
food additives, 190, 191
food adulteration, vii, viii, 187, 189, 190, 191, 193, 205
food industry, 197, 203
food safety, 187, 190, 194, 199, 203, 205
force, 21, 22, 25, 26, 31, 36, 89, 92, 94, 228
forensic nanotechnology, vii, viii, 26, 34
forensic science, vii, viii, 1, 5, 24, 25, 26, 31, 32, 33, 37, 38, 39, 41, 45, 49, 50, 55, 75, 76, 77, 78, 79, 80, 81, 82, 83, 96, 101, 106, 110, 112, 113, 114, 115, 121, 135, 145, 146, 147, 150, 174, 176, 177, 178, 179, 180, 182, 183, 184, 185, 216, 217, 224, 225, 227, 236, 253, 255, 282, 283, 291, 292
formation, 11, 37, 125, 133, 152, 160, 170, 274
fragments, 134, 148, 152, 153, 154, 155, 157, 159, 160, 182, 274

G

GABA$_A$ (modulator), 260, 261
GABA$_A$ receptor, 260, 261
gamma-aminobutyric acid (GABA), 259
gel, 37, 38, 149, 153, 157, 159, 161, 182, 221
gelation, 128
genes, 42, 104, 158, 170, 178, 180, 182, 210, 269
genetic information, 98, 173
genetic linkage, 158, 176
genetic predisposition, 171, 173
genetic testing, 172, 173, 185
genetics, 158, 174, 175, 179
genome, 145, 146, 147, 148, 154, 156, 157, 158, 159, 160, 166, 180, 183, 184
glucose, 60, 188, 264, 271, 275
glycine, 261, 264, 269
goat milk, 198, 209
gold nanoparticles (AuNPs), 36, 37, 38, 47, 57, 59, 65, 66, 78, 79, 81, 93, 95, 110, 111, 122, 124, 125, 128, 129, 133, 134, 136, 138, 167, 177, 178, 180, 182, 201, 202, 212, 213, 262, 266, 267, 268, 269, 270, 271, 274, 276, 282, 286, 287, 288
graphene oxide, 178, 182, 277, 288, 289
growth, 9, 36, 57, 65, 79, 81, 94, 146, 192
guanine, 122, 146
guilty, 148, 149, 169, 235
gun powder, 244

H

hair, 25, 41, 145, 146, 169, 185
handwriting, 146, 217, 238, 239, 240
haplotypes, 161, 182
harvesting, 192, 231
health, vii, 46, 52, 171, 172, 181, 187, 188, 190, 193, 205, 207, 209, 224, 248, 250, 275
health risk, 46, 187, 193, 207
heavy metals, 190, 192, 199, 200
height, 22, 164
hemoglobin, 26, 73, 100, 104, 105, 108, 113
heptane, 64, 69
heroin, 231, 233, 260, 261, 279
hexane, 64, 69
homeland security, 118, 143
home-made explosives, 117
human, 32, 41, 42, 45, 74, 75, 85, 101, 102, 103, 104, 105, 106, 107, 108, 112, 113, 147, 148, 154, 157, 158, 161, 166, 167, 170, 171, 175, 176, 177, 178, 179, 180, 181, 183, 189, 190, 193, 205, 211, 228, 236, 250, 265, 272
human body, 75, 101, 103, 112, 113
human genome, 147, 157, 158, 166, 178, 180
human health, 190, 193

humidity, 53, 127, 141
hybrid, 125, 139, 140
hybridization, 153, 177, 183, 198, 199, 202, 259
hydrazine, 71, 116
hydrogen, 6, 94, 95

I

identification, viii, 26, 28, 38, 39, 41, 42, 49, 51, 53, 55, 56, 85, 96, 97, 98, 100, 101, 102, 103, 104, 105, 106, 107, 108, 109, 110, 112, 113, 119, 135, 146, 147, 148, 149, 154, 158, 162, 163, 168, 171, 173, 175, 178, 183, 184, 194, 195, 197, 207, 210, 221, 223
identity, 19, 39, 118, 163, 166, 171, 172, 182, 194
illicit drug, vi, vii, 40, 255, 256, 258, 261, 275, 279, 282, 293
image, 17, 19, 21, 37, 38, 40, 50, 51, 57, 61, 62, 63, 67, 68, 70, 149, 218, 247, 274
immigration, 171, 178
immobilization, 183, 264, 270, 273
immunochromatographic, 106, 107, 108, 109, 114, 212
immunosurface, 122, 123
imprinting, 123, 128, 133
infrared (IR) spectroscopy, 101, 112, 197, 208, 257
inherited disorder, 145, 169, 170, 174
inhibitor, 106, 260, 261
injury, iv, 45, 245
intentional adulteration, 191
interference, 15, 50, 51, 88, 130, 262, 263, 264, 271, 276
iron, 61, 80, 189, 200, 205

J

judiciary, 244
jurisdiction, 243

L

latent fingerprint, viii, 25, 33, 39, 40, 49, 50, 51, 52, 53, 54, 55, 56, 57, 58, 59, 60, 61, 62, 63, 64, 65, 66, 69, 70, 72, 73, 74, 75, 76, 77, 78, 79, 80, 81, 82, 83, 274, 288
latent prints, 37, 53, 60, 76
law, vii, 15, 22, 135, 147, 172, 179, 180, 181, 184, 215, 227, 235, 241, 246, 247, 248, 249, 250, 253, 273
law enforcement, vii, 135, 235, 247, 248, 249, 250, 273
law enforcement agency, 247
laws, vii, 228, 243, 250
levorphanol, 261
light, 5, 16, 19, 23, 24, 28, 37, 39, 42, 57, 70, 72, 91, 92, 94, 98, 99, 108, 116, 146, 201, 220, 222, 223, 269
liquid chromatography, 120, 196, 207, 208, 266, 273
liquids, 12, 36, 196
liver, 189, 193
luminescence, 40, 54, 64, 66, 68, 71, 83, 119, 132, 222, 288
lysergic acid and its derivatives, 260

M

M2 (muscarinic) antagonist, 261
magnetic field, 167, 200, 276
magnetic nanoparticle, 32, 45, 183, 185, 200, 210, 264, 270, 285, 287, 290
magnetic resonance, 121, 257
magnitude, 86, 131, 134, 264
Marihuana, 261
mass spectrometry, 38, 40, 41, 120, 136, 197, 257, 266, 273, 275, 280, 290
materials, vii, viii, 3, 4, 5, 6, 8, 9, 10, 11, 12, 13, 14, 19, 23, 24, 30, 34, 35, 36, 37, 43,

44, 45, 52, 54, 73, 74, 75, 76, 80, 82, 86, 91, 96, 100, 107, 115, 116, 118, 120, 125, 200, 203, 206, 207, 210, 211, 213, 217, 221, 222, 224, 231, 242, 243, 244, 259
materials science, viii, 6, 19, 211
medical, 26, 27, 51, 53, 87, 121, 126, 147, 172, 173, 175, 177, 200
medicine, 5, 27, 31, 41, 82, 188, 203, 265
melamine, 187, 191, 193, 198, 203, 205, 208, 209, 213
metabolites, 40, 108, 257, 259
metal ions, 94, 134, 190
metal nanoparticles, 29, 91, 92, 121, 125, 127
metal oxides, 13, 14, 50, 136
metals, 10, 11, 14, 192
methadones, 261
methanol, 60, 66, 79, 131
methylphenidate, 261
microorganisms, 189, 190, 216
microsatellites, 157, 180
microscope, 4, 17, 19, 21, 29, 31, 99, 196, 221
microscopy, 18, 20, 25, 31, 78, 195, 210, 274
military, 117, 131, 142
miniature, 119, 135, 246
miniaturization, 168, 199, 255, 259
mitochondria, 17, 188
mitochondrial DNA, 26, 150, 176, 184
mobile phone, 35, 40, 45
mold, 194, 206
molecular biology, 109, 148, 175
molecular receptors, 259
molecular structure, 3, 36
molecular weight, 23, 269
molecules, 4, 6, 13, 23, 34, 35, 42, 44, 56, 59, 67, 86, 89, 90, 94, 117, 126, 128, 132, 134, 135, 139, 146, 154, 159, 161, 202
mono-amino oxidase inhibitors, 260
monoamino-oxidase, 260
morphine, 230, 231, 260, 261, 265, 266, 267, 268, 269, 270, 271, 272, 280, 281, 285, 286, 287
mRNA, 101, 102, 103, 109, 112, 113
mtDNA, 146, 150, 162, 163, 164, 174, 181
mtDNA analysis, 146, 174
multiwalled carbon nanotubes, 262, 273

N

Nafion, 127, 134, 268, 286
nano materials, 116
nano-based sensor, 33, 34
nanobiosensors, 104, 175, 293
nanocanals, 129, 142
nano-clusters, 35
nanocomposites, 45, 64, 67, 73, 80
nanocrystals, 5, 35, 45, 52, 64, 67, 80, 127, 131, 213
nanodevice, 42, 109, 259, 281, 293
nanodots, 28, 45
nanoelectronics, 5, 34
nano-enabled tools, 255, 256
nanofibers, 28, 45, 132
nanofibrous, 132, 142
nanoforensic, 146, 163
nanomaterial, viii, 1, 5, 8, 9, 14, 16, 24, 28, 29, 32, 38, 39, 46, 50, 69, 80, 86, 87, 94, 110, 115, 120, 121, 135, 187, 199, 204, 217, 219, 223, 224, 225, 259, 264, 281, 282, 284, 285, 293
nanomaterials, viii, 1, 5, 8, 9, 14, 16, 24, 28, 29, 38, 50, 69, 80, 86, 87, 94, 110, 115, 121, 135, 199, 205, 217, 219, 223, 224, 225, 259, 264
nanometer, 2, 3, 4, 6, 21, 35, 39, 86
nanometer scale, 2, 6, 35, 86
nanometers, 3, 17, 19, 22, 35, 200
nano-powders and nano-crystals, 35
nanoribbons, 89, 132, 142

nanosensor, 45, 85, 86, 87, 88, 89, 90, 91, 92, 93, 94, 96, 104, 110, 111, 112, 136, 140, 187, 200, 204, 211, 258, 259, 272, 276, 281, 287, 288
nanostructures, 8, 9, 34, 82, 93, 201
nanotechnology, vii, viii, 1, 2, 3, 4, 5, 24, 25, 26, 27, 28, 29, 31, 32, 33, 34, 35, 37, 38, 40, 42, 43, 44, 45, 46, 47, 48, 49, 50, 51, 54, 55, 73, 74, 75, 79, 81, 85, 86, 111, 115, 116, 120, 135, 142, 143, 166, 167, 177, 180, 187, 199, 203,210, 211, 212, 215, 216, 217, 218, 219, 220, 221, 222, 223, 224, 225, 227, 246, 248, 249, 250, 255, 259, 262, 264, 266, 281, 284
nanotube, 4, 44, 90, 91, 125, 126, 128, 131, 132, 134, 139, 140, 142, 143, 267
nanowires, 5, 29, 43, 45, 83, 120, 125, 127, 141, 200, 211
narcotic, 228, 230, 256, 260, 261, 266
national security, 87, 221
nerve, 260
nicotine, 41, 47, 260, 261
nicotine and cotinine, 41, 47
nicotinic ach, 261
nitramines or nitrosamines, 116
nitroaromatic, 116, 125, 130, 132, 134, 136, 137, 138, 139, 141, 142, 143
NMDA (antagonist), 260, 261
NMDA receptor, 260, 261
nuclear magnetic resonance, 197
nucleic acid, 88, 89, 151
nucleotides, 156, 157, 158, 161, 166

O

offenders, 161, 235, 242
Official Gazette, 230, 231
oil, 193, 196, 197, 207, 208, 229
olive oil, 196, 207
opinion evidence, 239
opium derivative, 230, 231
optical properties, 40, 51, 72, 94
optimization, 50, 52, 73, 74, 180, 224, 258
organ, 147, 174, 192, 272
organic compounds, 53, 68, 117, 191, 196
organism, 158, 198
oscillation, 29, 201
oxidation, 29, 56, 68, 73, 95, 202, 218, 262, 264, 267, 268, 272
oxide nanoparticles, 62, 80, 200

P

pain, 261, 265
pairing, 127, 128
para-magnetism, 121
parentage, 150, 158, 169, 170
parentage analysis, 158, 169
partial agonist, 260
pathogens, 43, 187, 199, 200, 204, 206, 210, 213
PCR, 26, 32, 101, 103, 104, 146, 149, 150, 153, 154, 155, 156, 158, 161, 163, 164, 166, 167, 176, 177, 178, 180, 182, 185, 198, 199, 201, 202
pentaerythritol tetranitrate (PETN), 116
peptide, 97, 106, 126, 129, 225
peroxide, 119, 131, 134, 142, 143
perylene, 81, 132
pesticide, 203, 209
petroleum, 28, 65, 223
pH, 56, 57, 59, 60, 105, 111, 124, 220, 224, 263, 265, 268
pharmaceutical, 44, 175, 209, 262, 264, 265, 272
phencyclidine, 260
phenylpiperadines, 261
phosphate, 69, 70, 83, 104, 105
phosphodiesterase inhibitor, 260, 261
photodecompose, 134
photoluminescence, 76, 127, 142

physical, 1, 6, 8, 12, 19, 30, 31, 37, 38, 49, 51, 53, 54, 55, 57, 59, 60, 68, 73, 76, 78, 83, 85, 86, 87, 89, 90, 92, 99, 116, 139, 165, 166, 188, 194, 195, 196, 198, 212, 215, 216, 217, 219, 220, 223, 227, 233, 255, 261, 292
physics, vii, viii, 4, 16, 86, 218
picrotoxin, 261
plants, 184, 190, 210
plasmonic, 92, 94, 110, 111, 201, 212, 284
platform, viii, 14, 32, 102, 120, 134, 167, 174, 175
polariton, 121
police, 78, 79, 234, 236, 247, 248
pollen, 196, 207
pollution, 19, 193
polymer, 23, 39, 69, 75, 76, 126, 128, 129, 141, 211
polymerase, 26, 154, 177, 182, 184, 185, 202, 209
polymerase chain reaction, 26, 154, 177, 182, 184, 185, 202, 209
polymerization, 154, 183
polymorphism, 150, 152, 155, 156, 157, 160, 161, 164, 165, 166, 176, 181, 182
polypropylene, 57, 65, 71
polystyrene, 58, 65, 132
poppy straw, 230, 231, 265
porosity, 220, 224
precipitation, 11, 60, 65, 262
premoline, 261
preservation, vii, viii, 216, 217, 218, 219, 220, 224
prisoners, 28, 222
probability, vii, 165, 174, 176, 182, 219
probe, 19, 21, 22, 44, 111, 121, 122, 133, 137, 202, 212, 262, 270, 273, 274, 275
proline, 196, 208
protection, 118, 174, 209, 213, 224, 250
proteins, 64, 101, 108, 126, 146, 158
pyrolysis, 13, 14

Q

quantification, 26, 27, 192, 208, 209, 262
quantum dots, 5, 13, 37, 38, 42, 45, 50, 51, 72, 77, 82, 83, 110, 134, 200, 203, 211, 213, 222, 271, 272, 287
quartz, 24, 125
quenching, 122, 123, 127, 130, 131, 132, 139, 142, 272
questioned document, 44

R

radiation, 15, 20, 44
radioactive, 44
Raman spectroscopy, 32, 98, 99, 112, 119, 120, 138, 143, 225, 257
RAPD, 146, 150, 156, 174, 180, 207
RDX, 116, 129, 133, 141, 143
reagents, 27, 31, 37, 45, 61, 67, 74, 77, 99, 264
recognition, 42, 86, 88, 90, 91, 94, 103, 123, 128, 164, 259
recovery, 38, 233, 234
reproduction, 146, 170, 240
residue, 37, 39, 40, 49, 51, 52, 55, 56, 57, 58, 61, 62, 63, 64, 72, 73, 74, 118, 189, 210, 218, 231
resistance, 56, 125, 274
resolution, 17, 19, 21, 44, 274
response, 121, 122, 127, 131, 263, 267, 268, 271
restoration, 219, 220, 225
restriction enzyme, 152, 159
restriction fragment length polymorphis, 149, 150, 154, 176
RFLP, 146, 149, 150, 151, 152, 153, 154, 155, 162, 174, 176, 182
RNA, 97, 101, 103, 112, 151
RNA and DNA, 101
room temperature, 68, 70, 117, 128

RSID-saliva, 107
RSID™, 105, 109, 114

S

saliva, 85, 96, 97, 98, 99, 100, 101, 102, 103, 104, 106, 109, 113, 114, 169, 257, 259, 269, 278, 280, 289
salmonella, 182, 202, 203, 212, 213
salts, 116, 229, 231
scaffold, 132
scanning tunneling microscope, 4, 21
scattering, 14, 17, 93, 125, 143
science, vii, viii, 1, 2, 3, 5, 24, 25, 26, 32, 33, 35, 37, 38, 39, 41, 45, 47, 49, 50, 55, 75, 76, 77, 78, 83, 86, 96, 101, 106, 110, 112, 113, 121, 135, 146, 147, 150, 166, 174, 178, 183, 184, 185, 207, 216, 217, 224, 225, 227, 259
scope, 243, 244
screen-printed carbon electrode, 262, 274, 285
secretion, 38, 51, 53, 101, 102, 104, 107
security, viii, 28, 33, 37, 38, 42, 43, 44, 49, 55, 118, 119, 120, 215, 216, 217, 221, 222, 224, 249, 250
security documents, 33, 37, 44
selectivity, ix, xii, 25, 49, 52, 56, 59, 72, 107, 111, 120, 122, 126, 129, 132, 255, 257, 258, 259, 263, 264, 273, 275
semen, x, 85, 96, 97, 98, 99, 100, 101, 102, 103, 104, 105, 107, 113, 149, 169, 257
sensing, vi, 22, 88, 91, 94, 110, 111, 120, 122, 126, 132, 136, 139, 142, 183, 195, 200, 202, 212, 255, 258, 266, 282, 283, 284, 285, 293
sensitivity, ix, xii, 5, 25, 43, 49, 51, 56, 59, 61, 66, 69, 72, 86, 96, 106, 107, 109, 120, 122, 123, 127, 132, 134, 141, 193, 198, 202, 203, 204, 212, 255, 257, 258, 259, 262, 264, 268, 273
sensor, ix, 5, 19, 29, 33, 34, 42, 44, 48, 55, 86, 87, 88, 89, 90, 91, 112, 116, 117, 118, 119, 120, 121, 122, 125, 126, 127, 128, 129, 130, 131, 132, 138, 140, 142, 143, 199, 204, 212, 246, 249, 257, 263, 265, 267, 268, 270, 271, 272, 273, 277, 280, 282, 284, 286, 287, 289
sequencing, 102, 112, 134, 147, 159, 178, 180, 184, 198, 199
serotonin, 260
serotonin re-uptake inhibitors, 260
serum, 106, 259, 262, 265, 270, 273, 274, 275
sex, 51, 53, 161, 162, 170, 172
silafluorene– fluorene, 128, 129
silane, 127, 128, 129
silica, 40, 41, 43, 47, 81, 82, 122, 123, 126, 128, 129, 133, 137, 141, 143, 201, 273, 275, 276
silica particles, 41, 47, 276
silicon, 13, 22, 30, 51, 69, 70, 83, 123, 265
silver, x, 25, 30, 37, 50, 51, 57, 58, 60, 65,
size–dependent, 93, 94
skeletal remains, 167, 175, 180
skin, xi, 31, 51, 80, 102, 103, 145, 162, 185, 188, 193
skin diseases, 188, 193
smokers, 41, 47, 279
smoking, 228, 231
sniffer, 118
sodium, 63, 64, 69, 109, 268
sodium dodecyl sulfate (SDS), 63, 268
species, 88, 89, 126, 128, 132, 147, 158, 201, 208, 229, 231, 257, 263
specimen, 17, 18, 22, 163, 165
spectroscopic techniques, 98, 119, 257
spectroscopy, 14, 23, 24, 31, 32, 98, 101, 112, 119, 120, 138, 139, 143, 196, 208, 225, 257, 266, 272, 275
STR, 80, 107, 146, 150, 157, 158, 161, 162, 163, 164, 166, 172, 174, 175, 177, 178, 180, 182, 184

strychnine, 261
surface modification, 51, 61, 72, 124
surface plasmon, 38, 91, 92, 93, 121, 124, 133, 137, 138, 143, 200, 201, 211, 212, 266, 276, 279
surface plasmon resonance, 38, 91, 92, 93, 124, 133, 137, 138, 143, 200, 201, 212, 266, 276, 279
surfactant, 36, 57, 61, 63, 220
surveillance, 135, 247, 249
sweat, 25, 51, 53, 68, 96, 97, 98, 99, 100, 104, 109, 179
synthesis, viii, 1, 5, 8, 9, 11, 30, 78, 82, 83, 86, 140, 293

T

tandem repeats, 147, 150, 157, 161, 178, 180, 184
target, 11, 40, 43, 44, 68, 88, 90, 92, 94, 102, 103, 117, 118, 120, 128, 158, 200, 201, 202, 203, 267, 273, 274, 276
temperature, 12, 13, 29, 53, 56, 90, 117, 123, 128, 153, 154, 157, 189, 210
terrorism, 43, 117, 118
terrorist activities, vii, 115, 242
THC, 261, 280
The Narcotic Drugs and Psychotropic Substance Act, 228
thin films, 5, 12, 15, 30
tin, 205, 266
tissue, 88, 102, 103, 147
titanium, 13, 79, 131, 134, 142, 143, 167, 180
toluene, 126, 129
Touch DNA, 146, 162, 184, 185
toxicity, 46, 72, 137, 183, 265
toxicology, viii, 206
transducer, 88, 90, 91
transduction, 90, 91, 258, 259
transistor, 90, 92, 125
transistors, 126
tricyclic anti-depressants, 260

U

unintentional adulteration, 191
urea, 68, 108, 193, 194
uric acid, 263, 268
urine, 27, 32, 96, 97, 98, 99, 100, 104, 106, 107, 108, 257, 262, 265, 267, 268, 269, 272, 278, 280, 289
urine and sweat, 96, 97

V

vaginal fluid, 96, 97, 98, 107
vapor, 12, 30, 68, 117, 119, 120, 123, 125, 128, 134, 141
vegetables, 191, 194, 206, 210
victims, 162, 174, 242
violence, 118, 246
violent crime, 99, 100
viruses, 91, 92, 140
vision, 50, 111, 193
visualization, 38, 53, 60, 61, 62, 66, 70, 73, 76, 79, 81, 100, 137, 202
volatile organic compounds, 117, 196

X

xanthines, 261
X-ray diffraction, 14, 15, 30

Z

zinc, 25, 39, 40, 44, 63, 66, 71, 79, 134
zinc oxide (ZnO), 39, 44, 52, 66, 76, 79, 139, 175